A

Commercial Applications of Wind Power

Commercial Applications of Wind Power

Paul N. Vosburgh
Albuquerque, NM

VNR VAN NOSTRAND REINHOLD COMPANY
NEW YORK   CINCINNATI   TORONTO   LONDON   MELBOURNE

Library of Congress Catalog Card Number: 82-17361
ISBN: 0-442-29036-5

Manufactured in the United States of America

Published by Van Nostrand Reinhold Company Inc.
135 West 50th Street, New York, N.Y. 10020

Van Nostrand Reinhold Publishing
1410 Birchmount Road
Scarborough, Ontario M1P 2E7, Canada

Van Nostrand Reinhold
480 Latrobe Street
Melbourne, Victoria 3000, Australia

Van Nostrand Reinhold Company Limited
Molly Millars Lane
Wokingham, Berkshire, England

15 14 13 12 11 10 9 8 7 6 5 4 3 2 1

Library of Congress Cataloging in Publication Data

Vosburgh, Paul N.
    Commercial applications of wind power.

    Includes index.
    1. Wind power.   I. Title.
TK1541.V67   1983   333.9'2   82-17361
ISBN 0-442-29036-5

To Jane Berry Vosburgh for the support, encouragement, editing, and indexing that made this book possible. To Susan, Kay and Junior, the benefactors of future wind power. To all those who are pioneering wind power now. Thanks.

Introduction

Wind power is ready to be utilized now in many practical applications. It has long been considered a clean and renewable domestic source of energy and has been widely used for mechanical power and battery charging. In recent years, oil embargos, environmental concerns, political tampering, and rapidly escalating costs have brought the need for alternative and renewable energy into the limelight. Of all the renewables, wind is clearly the closest to being ready to become part of the nation's energy mix.

This book discusses wind energy conversion systems (WECS) capable of generating substantial quantities of electricity, arbitrarily defined as WECS which utilize one or more wind turbines with generating capacity of 20 kW or more. It limits itself to commercially promising systems which generate utility quality power such as needed by most homes, farms, and businesses. Very small WECS, and WECS intended for mechanical power or specialty applications, are specifically excluded.

Modern WECS of substance have had to overcome the belief that converting raw windpower to useful electricity is easy and should be cheap. After all, the power of the wind has been used for thousands of years to perform useful work. However, turning that windpower into electricity is much more technically demanding and complex than using it directly for sailing ships, grinding grain, pumping water, or performing other mechanical tasks. Generating useful alternating current (AC) has also proven to be much more demanding than direct current (DC), which was important and common in the early twentieth century before utility power lines were extended to almost every home and farm in America.

None of the new WECS has had operating experience of more than a few years. Many have had problems, some major and some very dramatic. Most of the WECS machines and projects now up have to be considered research or demonstration facilities. Even the strongest WECS advocates now acknowledge that the only credible test of a WECS concept or design is the test of time applied to full scale hardware installed in real world conditions. The time for such a test to be meaningful is not universally agreed on, but most participants are looking for evidence that the installed WECS can operate reliably and safely without unusual or costly repair and maintenance for a period of at least twenty years. At least 100 hours of wind driven generation is generally required before the WECS is even accepted for automatic operation. Most utility planners want to see at least two or three years of successful operation before they have confidence to specify WECS in their generation systems.

Prior to 1980, only a few pioneers were proceeding toward commercialization of WECS without direct government assistance. And even they were helped by a generally positive attitude by governments at all levels and by liberal tax incentives and pressure on utilities to cooperate. Those fortunate enough to receive direct government aid got it in the form of research and development funding, direct government purchases for demonstration projects, and promotional and marketing assistance. Some of the private and publicly supported efforts led to encouraging installations. All had at least minor problems. And many failed dramatically by throwing components or completely self-destructing. Even successful prototypes were modified, or served as the basis for improvements and changes in later models, so that actual testing time of today's commercial offerings is extremely limited. In 1982, not one WECS producer can claim more than a year of successful operating experience with specific hardware being offered for sale.

At best, the emerging WECS industry is in its embryonic stage of evolution. Some authorities argue that there is not yet an industry, or perhaps one in the pre-embryonic stage. In any event, it is not well established. It is in the process of shaking itself out and responding to birth and growing pains which accompany any major innovation.

This book concentrates on the realities of today's WECS and practical applications for the near future. It doesn't attempt to present the fine points of the technology or show you how to design and build your own wind machine. Several earlier books by others have covered those topics in great detail. Conversion of wind to electricity has, in fact, been demonstrated. The trick now is to do it in an affordable way that is safe and reliable over enough years to make substantial investments in the process attractive.

WECS have evolved a vocabulary of their own and, probably because of the massive government involvement, have introduced many acronyms into the language. Appendix B attempts to provide definitions and clarify acronyms. However, if you are relatively new to the subject, be prepared for liberal use of some new words. *WECS* itself is utilized throughout as both the singular and plural for one or more complete wind energy conservation system. Occasionally it is used to describe only the wind turbine generator. *DOE* is, of course, the United States Department of Energy, which also sometimes is credited with activities of its predecessor agency, Energy Research and Development Administration (ERDA). *Power* is the instantaneous measure of work being performed by a WECS, either expressed in mechanical terms as horsepower (hp) or in electrical terms as kilowatts (kW) or megawatts (MW). One hp is equivalent to a little less than 0.75 kW. *Energy* is power over time and is commonly expressed as kilowatt hours (kWh) or megawatt hours (MWh). *Velocity* is usually expressed in miles per hour (mph), metres per second (m/s), or revolutions per minute (rpm).

This book doesn't dwell on future goals or projections for energy contributions by WECS. There is enough challenge in figuring out what is going on today. However, the official United States goal is to have WECS provide approximately 2% (equivalent to about 1.7 quadrillion British thermal units, BTUs) of the nation's energy by the turn of the century. That relates to almost 10% of America's electrical demand and could save 1.5 billion barrels of oil annually. Some big thinkers like to put the word "only" in front of the phrase "2% (1.7 quads) of the nation's energy" and dismiss the potential as insignificant. For perspective, that total is equivalent to:

- Enough energy to heat 850 thousand typical homes for 20 years.
- Enough crude oil to fill a 127 ship fleet of 325 thousand ton supertankers with capacities of 2.3 million barrels. (This quantity is tripled if the 33% efficiency of converting oil to electricity is factored.)
- A pile of coal (approximately 70 million tons) large enough to fill 850 thousand railroad cars.
- 500 billion kWh of electrical energy (based on 3412 BTU/kWh).

The government goal is probably not achievable. With the oil glut and recession of 1982 many government leaders and others have given up on it. But even much less of a contribution can be both helpful to the nation and enormously profitable to the suppliers and owner-operators of WECS. Although a few billion dollars doen't mean much to the federal bureaucracy, it should be enough to power the pioneering efforts needed to move WECS from the R&D stage to a point where they can become significant contributors to the nation's energy mix while providing reasonable financial returns to those willing to invest in them. This book is intended to provide guidance toward that modest goal.

Contents

Commercial Applications of Wind Power

1
Growing Pains:
An Embryonic Industry

The emerging wind energy conversion systems (WECS) industry has been hampered by the assumption that building and using modern WECS is easy. It isn't.

Because windmills date back to before Christ, people were quick to accept the idea that the wind can be readily captured and turned to useful energy. It can, but not everywhere and not cheaply.

Many people also jumped to the conclusion that because no one controls the wind, that energy from the wind is available to them at no cost. Available? Probably. No cost? No way.

On the positive side, wind is available in many places in North America in sufficient quantity to make it worth capturing. The wind itself is locally available and, therefore, doesn't have to be imported from foreign countries. It is also difficult for governments to tax or ration, or for some corporate giant to monopolize. The only trick is to collect it when it is available, convert it to something useful, and be able to use it when it is needed.

In addition to being relatively locally available (over a small percentage of the continent's land) and free (in its raw form), wind is nonpolluting (if you don't count the dirt, sand, dust, salt spray, leaves, branches, tumbleweed, and other debris that it sends along), nondepletable (yes, a renewable raw energy fuel dependent only on the Sun and the Earth's rotation), and it doesn't have to be dug or pumped out of the ground or trucked or otherwise transported to its point of use.

Some advocates also claim "environmentally benign" and "safe" as inherent advantages of wind in comparison to other alternatives. In general, that's true. However, it is difficult to think of tornadoes, hurricanes and other storms as either benign or safe. Wind storms cause more than their share of death and destruction in any given year. Wind is a potentially treacherous force which must be treated with great respect by those who want to work with it.

An industry is starting to evolve to work with the wind and turn it into energy useful to all. It is not quite an industry yet; more a loose confederation of advocates and people and companies seeking a meaningful role for themselves and for WECS. In strategic planning terms it might be advanced enough to be called an embryonic industry. However, some may argue that it is pre-embryonic, or just a gleam in the eyes of the advocates and innovators.

Current WECS activity is dominated by research, development, and demonstration and characterized by trial and error efforts. There have been more failures than successes. More companies have gone out of business than are still in. Hopefully, those misfortunes were only healthy growing pains which have built a solid foundation for responsible WECS contributions in the future.

WIND TO ELECTRICITY

Converting wind to electricity is the name of the game. And that requires a complex interaction between the wind, the ground, a means to capture, a mechanism to convert, a method of transforming the energy into a useful form, and then transmitting it to the place where it is to be utilized. The total process is the wind energy conversion system. All the many parts must work as one. The total system is only as good as its weakest link.

The wind can be captured and converted to useful mechanical energy, such as for pumping. The traditional American windpumpers were, and are, good examples of this reliable approach. Or, wind can be captured and converted to heat, as has been demonstrated in water heating applications. But those applications, as well as the very small electricity generating WECS and the battery charging direct current (DC) windchargers, are not the subject of this book.

The topic here is converting wind to the type of electricity used in America, high quality alternating current (AC) compatible with the nation's in-place generation, transmission, distribution, and utilization systems. Even if a WECS is not directly tied to a central electric system, or grid, its output must be utilized in appliances, tools, or machines generally built to work with utility quality electricity.

By eliminating all the WECS variations which don't result in utility quality electricity (and the very small WECS even if they do), this book has been kept much thinner than it might have been. But the WECS remaining are the ones that are going to be the backbone of the WECS industry and the ones that will make a substantial contribution to the nation's energy supply by favorably competing economically with established generating systems and their fuels, and with other emerging alternatives.

Within a total WECS, the *wind* itself is the most important single ingredient. Without powerful winds available when needed there is no chance for the other components of the system to function in an economically attractive way. In conversion formulae, wind velocities are cubed while other numbers are linear. If strong winds are not available, WECS cannot compete with established generators and their fuels. As obvious as that conclusion should be, it has been ignored over and over again. In the United States, if you don't have access to a site which is swept by winds averaging at least 14 mph, measured at approximately 30 ft above grade, don't plan on finding a WECS which can be cost competitive. (You may want to build one anyway, but it should be for other reasons.) In areas with low competing energy or fuel costs (or prices), windspeeds have to average even higher. And the averages must be made up of a good mix of windspeeds. With the cubing factor, a 14 mph average made up of half zero and half 28 mph is dramatically better than a 14 mph average based on constant 14 mph breezes. And watch out for the high storm winds, or excessive turbulance, or good winds at the wrong time of day, or day of week, or time of year, etc., etc. If sufficient quantity and quality of windpower is available, you've got a good start toward a worthwhile WECS.

Probably second in importance in a good WECS is the *ground.* A site must be available for purchase or use. It must be affordable in cost. It has to be capable of supporting the structural loads and being built upon. Current or future protrusions can't obstruct the wind. It must be accessible for delivery, construction, and operating and maintenance activities. If not where the electrical output is to be utilized, it should have access to a powerline or substation. And political, legal, and public interests will have to allow use of the ground for WECS. Without a place to structurally and electrically interconnect hardware there can be no WECS.

The WECS hardware itself is discussed in more detail in Chapter 2, but regardless of type, size, or details, the two major functions of the hardware are to efficiently *capture* the kinetic energy of the wind and to efficiently, reliably, and safely *convert* that energy to electricity ready to use or transmit.

The *capture* function is universally accomplished by means of rotating blades attached either to a hub (horizontal axis wind turbines, HAWT, or propeller types) or a central rotating tower or torque tube (straight or curved blade vertical axis wind turbine, VAWT, or egg-beater types). The amount of wind useful to the WECS is limited to the swept area of rotor which the wind passes through. Swept areas, in turn, are defined by HAWT blade lengths (radius of a circle), straight blade VAWT blade lengths and rotating diameter (a rectangle), and curved blade VAWT rotor height and diameter (approximately and ellipse defined by the blade curvature). HAWT rotors can be located either upwind or downwind from their support towers and are moved (yawed) to face directly into or away from the wind for efficient capture. HAWTs and straight blade VAWTs generally move (pitch) their blades to provide favorable angles of contact with the wind. The structural towers, or supporting base assemblies, for the rotors contribute to the capture by raising the rotors off the ground and into usually higher winds. Shape, size,and number of blades is generally based on a combination of aerodynamic efficiency, structural capability, and cost. The rotors can spin at constant speed (chosen to be most efficient at one target windspeed) or at multiple or variable speeds (to be most efficient at more than one windspeed). All these variables combine to define the amount of available wind captured in the form of rotating shaft torque.

After the rotor has captured the kinetic wind energy and turned it into rotating mechanical power, a drive train speeds up the rotation and transmits the torque to the generator to *convert* the mechanical energy to electrical energy. The drive train usually includes a gear box or speed increaser, rigid and flexible couplings, diagnostic sensors, emergency brakes, and other features for environmental protection and safety. The generator is usually either synchronous or induction for direct AC electricity, or a DC generator for indirect AC electricity. Everything is interrelated by sophisticated controls which also integrate the WECS hardware with the wind (usually by means of an anemometer) and the user load or utility grid.

After conversion, the WECS must *transform* the raw electrical energy into the proper quality electricity for use or transmission. If the WECS hardware generates DC, an inverter has to bring it to AC. If the line is too weak for the WECS generator, capacitors may have to be added for power factor improvement. If generation voltage is not what is needed by the user (or in the grid's transmission or distribution line), a transformer or substation will be needed to bring it up or down to the appropriate line voltage.

The final link in the total WECS is the means to *transmit* the electricity to where it has the most value. If it is to be used on site, it doesn't have to move far but it needs the proper kind of interconnections with the load or service panel. If the electricity is to be integrated into a utility grid, it must be properly interconnected with the available transmission or distribution line. In all cases, safety, lightning and ground fault protection, ability to disconnect for service, metering, and control are important features of the total system.

THE STATE OF THE ART

The process of turning wind into electricity is fairly well understood, and many inventors and organizations have contributed to the growing base of technology. However, the state of the art is still in its infancy. The main problem is that it is extremely costly to demonstrate or test new concepts. Although much can be modeled and analyzed by computer, or on paper, proof-of-concept only occurs when a full scale prototype demonstrates success under real world conditions.

When interest in wind energy was rekindled in the '70s (after the first OPEC oil embargo) numerous proposals were offered for new and improved WECS and bigger and better models. On paper, and at the proposal stage, many concepts held out the promise of low cost generation of electricity, and many still do. But for one reason or another, as WECS moved from the concept stage through design and engineering and into the prototype stage, almost all the solutions got more complicated and, in turn, more expensive. More important, most did not pass the ultimate test, that of survival or trouble-free operation.

So the shake out began in the late '70s and is still in progress. Major multinational companies have been in and out of the industry. Many special purpose WECS companies were started up, and became visible enough to be noticed, and then vanished when the problems and costs started to mount. Still others tried but had to give up before they even got to the start-up phase. Most of the survivors (see Chapter 2 for the producers of electricity machines) have had major problems and delays and setbacks in their programs.

So the state of the art is the technology and WECS hardware that has survived the test of installation and operation. As Alcoa's wind energy venture manager, Frank Townsend, explained to the audience at California's 1981 conference on wind energy opportunities (right after Alcoa's world's largest VAWT self-destructed in the Palm Springs Desert as a feature of that conference), "Working on WECS is like performing R&D in a fishbowl. There is no way to innovate with the technology unless you're willing to expose yourself to public view." And as Alcoa and dozens of others found from experience, there is no such thing as a small problem, a simple solution, or a low cost fix. Although initially underestimated by practically everyone who has entered the field, WECS technology is among the most complex and technically demanding of energy systems.

WECS combine all the technical demands of complex rotating electromechanical equipment with the architectural and engineering demands of a building or free standing structure. In addition, wind and other environmental conditions bring into play all the meteorological vagaries and "acts of God." If that's not challenging enough to scare a reasonable person, add the intrigue of government and political tampering and throw in the goals of low cost and maintenance free life expectancy of 20 years.

So it is not surprising that current WECS efforts are concentrated on incremental improvements to proven technology. Innovation rather than invention. Evolution instead of a breakthrough.

Most current U.S. HAWT efforts can trace their roots to the early Smith-Putnam, Gedser, MOD-0 and smaller prototypes (see Appendix A—WECS Historical Evolution). Current VAWT efforts are directly descended from Georges Darrieus' inventions in the '20s, reinvention at the National Research Council of Canada in the '60s, and refinement in the '70s at both NRC and Sandia National Laboratories. Government and private efforts are even starting to reference their designs as second (NASA's MOD-2 and Sandia's low cost 17 metre VAWT, for instance) and third (such as NASA's MOD-5s, Westinghouse's WWG 0500, and the Flow, Forecast and DAF VAWTs) generation designs. The emphasis is properly on learning from and building on past experience.

Although there have been many attempts at other types of WECS innovation, at the time of this writing the state of the art is limited to two and three bladed upwind and downwind HAWTs and two and three bladed Darrieus type VAWTs. Those probably aren't the ultimate answers, and might not even be the best choices today, but they represent all there is—today's reality.

THE HELPERS

The embryonic WECS industry has had a great deal of outside assistance and many of the helpers deserve recognition, as well as an appeal for further help as the industry starts to take shape.

The American Wind Energy Association (AWEA) certainly has to be counted as one of the prime movers. They led the lobbying for government R&D funding, and PURPA, and tax credits, and anything and everything else believed helpful to WECS. AWEA's national conferences have been the focal points for information dissemination and business deals that served to get most of today's participants started. Although AWEA was usually controversial (and always knocking at bankruptcy's door) it has been the industry's only cohesive force. Credit should be given to AWEA's pioneering leaders back in the ancient '70s. Rick Katzenberg (Natural Power), Don Mayer (North Wind Power), Herman Drees (Pinson Energy), Vaughn Nelson (Alternative Energy Institute), Steve Blake (Sunflower Power), and Bill Batesole (Kaman Aerospace) were the willing workers I recall as tireless directors trying to build something out of nothing. And their always enthusiastic and optimistic executive director was Ben Wolff, who established a Washington presence for the industry. AWEA has started to mature and move toward a more established trade association under the current leadership of executive director Tom Gray and Directors representing both the pioneering smaller companies and the major corporations (General Electric, Westinghouse, Alcoa, Windfarms Ltd., etc.) who later entered (and in some cases have already left) the industry. If you haven't already joined, do so. It's cheap.

The U.S. Department of Energy (DOE) has tried to help with all aspects of the WECS industry. Unfortunately, continuity and consistency are not part of the federal vocabulary. Congress mandated conflicting and counterproductive goals to DOE. The DOE secretaries under four presidents have had only limited success in getting serious attention for WECS. The 1982 secretary of the day, Jim Edwards, is dedicated to dismantling DOE entirely and moving into the world of academia. The normal government bureaucracy has moved at its own slow pace while imposing its own obstacles to progress. Through it all, the DOE wind program manager, Lou Divone, has kept his sanity and, miraculously, has gotten some winds measured and mapped and 15 WECS of substance (four MOD-0A HAWTs, three low cost 17 metre VAWTs, one MOD-1 HAWT, five MOD-2 HAWTs, and two 40 kW SWECS) up and into demonstration and test modes of operation. Although none of the DOE research prototypes has yet been turned into commercially available hardware, there is encouraging movement toward bringing commercial versions of DOE's MOD-0A HAWT and low cost 17 metre VAWT, as well as uprated versions of the two DOE 15 kW HAWTs developed and prototyped at their Rocky Flats test center, into the market place by 1983.

Elsewhere in the federal government, Dr. R. Nolan Clark's applications research program for the U.S. Department of Agriculture (USDA) at their Bushland, Texas wind research station has been a most productive assist for WECS up to 100 kW in size intended for agribusiness and irrigation installation. At last count, USDA was testing nine WECS at Bushland. (See Chapter 4, Dispersed Applications, for more on the USDA programs).

The major state WECS assist programs (in California, Hawaii, and Oregon) and some of the key helpers are also discussed in Chapter 4, in the section on Governments and the Institutions. In addition to those major programs, many other states have provided either formal or informal help to the emerging WECS industry. State officials in

Wisconsin cooperated with Wisconsin Power and Light and their Dr. Carel De Winkel in an ambitious dispersed small WECS demonstration program and innovative rate structure. New York State has actively prospected for wind resources and cosponsored a few small WECS demonstrations. New Mexico has sponsored work at its two major universities that led to WECS courses and training programs, as well as wind mapping, under the leadership of Dr. Ken Barnett at New Mexico State and some demonstrations and wind farm studies under Dr. Gerry Leigh at University of New Mexico. Texas has sponsored much of Dr. Vaughn Nelson's impressive work at the Alternative Energy Institute, West Texas State University. In Pennsylvania, Paul Gipe's Center for Alternative Resources has been a catalyst for the commonwealth's WECS activity.

At one time, there were even Regional Solar Energy Centers (RESEC) which had modest wind energy assist staffs to serve as coordinators of regional state activities with the federal programs. Ross Bisplinghoff at the Northeast Solar Energy Center (NESEC) and Steve Nelson at Southern Solar Energy Center (SSEC) were especially effective in stirring up WECS activities in their regions.

In the utility sector, the National Rural Electric Cooperative Association (NRECA) has an active WECS research interest headed by Lowell Endahl and Wilson Prichett. American Public Power Association (APPA) wind activities have been led by Dr. R. Eric Leber, director of energy research. The Electric Power Research Institute (EPRI) has Drs. Ed DeMao and Frank Goodman as WECS advocates for the investor-owned utility industry. Excellent reports on many aspects of WECS, particularly as they relate to direct utility use or interaction with utility systems, are available, as is good practical advice, from all three organizations or any of those knowledgeable WECS specialists.

And helping to tie everything together, a few specialized publishers have dug into WECS and have done an outstanding job of keeping their readers up to date in a fast changing environment. In addition to the leading WECS book publisher, Van Nostrand Reinhold (of course), Farrell Smith Seiler's *Wind Energy Report* is always on top of the inside news on a monthly basis. *Solar Energy Intelligence Report* brings us WECS news every week. Don Marier's *Alternative Sources of Energy* is coming on strong. And a new major force is

Renewable Energy News, which started out as a Canadian magazine and now covers the world quite well. Mike Evan's *Wind Energy Digest* was the voice of the industry in the '70s and Mike was also one of the pushers for a responsible and broad-based wind energy association.

The point of all this is that a WECS industry is starting to take shape. In some windy areas of the country all the ingredients are in place for rapid industry growth and installation of hundreds of WECS projects. All that is needed is a little leadership, a little money, and a lot of hard work.

OPPORTUNITIES ARE AWAITING

Opportunities are awaiting for everyone. Good reliable WECS hardware can be sold in quantity. Utilities can install fleets of wind turbines to help reduce dependence on depletable oil, coal, water, and uranium fuels. Entrepreneurs can organize land, labor, capital and management and build profitable wind farms. Farms, commerical and industrial facilities, major processors and governments and institutions that have access to windswept land can start saving on fuel bills or harvesting the wind for sale of electricity. Those major opportunities are discussed in more detail in Chapters 2, 3, 4, and 5. In addition, there are almost unlimited personal opportunities for individuals, professionals, investors, landowners, lenders, insurers, truckers, contractors, suppliers, promoters, sellers, packagers, etc.— what the government and AWEA's Ben Wolff used to call the industry infrastructure.

Just as in any established industry, before WECS can be widely used and make a significant contribution to the nation's energy mix, all the barriers must be overcome and a wide variety of constituencies must become involved. Inertia must be overcome and the bureaucracy has to have its procedures in place so that WECS can be treated in a routine way. Harry Cruver, a leading Virginia energy and management consultant, became perplexed at what he called "the prodigious size of the private and public international bureaucracy engaged in energy, including wind activities, on behalf of the developing nations." Harry's new company, Energy Applications Corporation, was attempting to establish WECS as a viable alternative in the windy Caribbean when the reality of inertia hit him. His observation that actual results, in terms of realistic or effective technology achievements, were minuscule relative to the attempts to help, is probably just as applicable to local WECS projects as in the developing nations market. Harry's observation that "wind power will come of age when the energy actually delivered to users equals the thermal energy that would be generated by burning all the reports and studies written on the subject" is my favorite.

There still is a need for consultants, writers, studiers, teachers, information disseminators, and organizers of workshops, conferences, and seminars. And all of those activities offer personal opportunities for people good at them.

But even greater opportunities are awaiting those who get involved in the mainstream of the WECS industry, in the producing; marketing, distributing, installing, and servicing; and financing and managing functions. And in the functions that can remove obstacles and make the total process work, such as participating in related technical societies, wind energy associations, organizations for promulgating and implementing codes and standards, zoning boards, energy policy bodies, insurance and financial institutions, and other regulatory or policy making agencies.

In the WECS mainstream, production needs research and development, engineering, procurement, fabricating, quality assurance, safety, and overall production management and accounting and control skills. Typical producers provide broad opportunities for component and materials suppliers, job opportunities for skilled and unskilled plant workers, and work for support and ancilliary industries (such as trucking, warehousing, and equipment supply).

Within the marketing, or market access, function, producers need their own marketing specialists, but they also need a myriad of independent sellers and providers of services to get the WECS hardware from the production line into service and paid for. There is almost a desperate need for providers of turnkey services who can provide all the functions to get the hardware sold, installed, started up, and kept in service. Together or separately, those opportunities include selling (manufacturer's representatives, commission salespeople, distributors, and dealers), applications engineering (wind prospecting, site analysis, feasibility studies, code and regulatory agency compliance, customizing, and generally providing sales support), site design and engineering (civil engineering, electrical engineering, construction engineering, site planning and surveying, meteorological engineering, and securing of permits and variances), site preparation (grading, roads, power lines, lightning protection, utility interface, foundations, anchors, fencing, etc.), and on-site assembly and erection of the WECS hardware (mechanical assembly, rigging, steel erection, mechanical and electric hook-up, environmental protection, inspection, and start-up and modification). Finally, there is an opportunity to provide warranty and operating and maintenance service.

The managing and financing functions are multifaceted in that they are required ingredients in all the other functions and also necessary in pulling together all the resources needed to get a WECS project conceived, planned, developed, and into successful operation. The potential owner-operator of the WECS must have equity and/or debt financing, as well as windswept land, before it can seriously consider being a customer for WECS hardware and services. Unless that owner-operator has had experience with WECS, there will be an opportunity for a project manager. Somebody has to organize and control all aspects of a project and keep the bankers, insurance companies, investors, lawyers, accountants, tax collectors, and other participants fully involved as needed.

All of these opportunities are greatest during the embryonic phase of WECS industry growth. A window of greatest opportunity will probably exist until about the early 1990s, perhaps as long as the turn of the century. By then, the industry will either be fully established and into its mature stage of growth, or will be back on hold for another generation.

One of the industry's leading prognosticators, Strategies Unlimited Incorporated, told its clients that "Today's electric utility industry does have sufficient excess generating capacity to survive any short-term peak requirements and general growth characteristics. However, within only a few short years, that excess capacity may very well be absorbed, since it is becoming extremely difficult to implement gas or coal fired plants, and especially nuclear powered plants. If this trend is continued, by the mid 1980s the U.S. will be faced with severe energy limitations and, depending upon the depth of the supply constrictions, reactive programs will be implemented."

They explained, "It will take time for these programs to result in the necessary generating capacity to overcome the deficit and cover the potential growth in both peak and base load requirements. The only possible reactive programs that can be implemented will be those based upon conventional fuel sources. Alternative energy sources will not be capable of meeting the necessary requirements in the electric utility industry. Once the reactive programs have been implemented, it will take no more than ten years to reestablish major expansions in conventional energy generating facilities."

Under the Strategies Unlimited scenario, alternative energy sources have no more than ten to fifteen years to establish themselves as credible energy sources. If the suppliers of equipment and technology delay their market development efforts until the late 1980s, it will be too late to establish alternative energy sources as a significant portion of the nation's generating capacity. Strategies projected that "The reactive programs that will have to be implemented will be even more dramatic than might otherwise be expected if the alternative sources cannot be established in the next five to, perhaps, eight years on a solid footing. Any capacity that can be supplied by the alternative sources will assist in delaying the need for implementation of reactive programs in the conventional energy domain."

During that window of opportunity, the American Wind Energy Association's objective of widespread utilization of wind energy will become increasingly important. AWEA conceptualized five stages of WECS utilization:

- *Knowledge.* Potential users must become aware of the new WECS products and related services, must gain an understanding of how they function, and must assess their potential impact on current practices.
- *Persuasion.* A set of attitudes must be developed that facilitates adding wind energy to the available energy options when new facilities are needed.
- *Decision.* Capabilities to make human, social, technical, and economic feasibility decisions that will facilitate those changes which must be made, and the nerve to move forward in the face of uncertainty.

- *Initial use.* Users must experiment with new WECS products or services on an evaluation basis to build up their confidence and understanding and allow them to make intelligent investment decisions.
- *Continued use.* The new technology must be integrated into the mainstream of current practice. WECS must move from the novelty, or R&D, stage into commercial use.

Combined with successful development efforts, those utilization factors should provide the opportunity for wind to achieve its potential as an inexhaustible, nonpolluting, domestic fuel for affordable electrical energy. Working toward, and fulfilling, that goal will provide opportunities for many people and companies, and for the two major beneficiaries, those who consume electricity and those who market electricity. Opportunities are awaiting.

2
WECS Technology:
The Hardware

The evolution of contemporary electricity generating WECS technology essentially started in 1973 as a direct result of the first OPEC oil embargo and resulting price increases. Since that time, many governments and private companies have worked to develop WECS hardware which could be utilized as alternatives to more established power plants.

The emphasis has been on equipment to efficiently capture and convert the intermittently available wind into useful electricity compatible with established energy generation, transmission, distribution, and utilization systems. There have been many failures and false starts, but progress in the past decade has been truly impressive. By the mid '80s, ten to 15 producers expect to be offering tested WECS capable of generating 20 kW to 4 MW of power in the United States. Intended utilization ranges from single unit dispersed applications to giant multi-unit windfarms integrated with utility generation systems.

The impressive technological evolution of WECS is discussed here as:

- Basic concepts and approaches;
- Technical risks and immaturity;
- Research and development thrust; and
- Producers of electricity machines.

BASIC CONCEPTS AND APPROACHES

Basic concepts and approaches to WECS hardware have been as numerous as the number of people who have worked on them. An almost universal problem has been underestimating the complexity of the

15

task. Probably because windmills and windpumpers have centuries of history (see Appendix A) and windchargers have been around since the 1890s, it was assumed easy to design and build a modern WECS. It wasn't. And it still isn't.

Developing hardware to effectively convert intermittent, variable, and often treacherous winds into the type of 60 Hz AC electricity Americans expect (and which their appliances and power systems are built for) is technically demanding by itself. Adding the variables of land, weather, laws, politics, and institutional inertia doesn't make it easier. And attempting to do it all at costs that can be competitive with established alternatives, or even be affordable, has tended to separate the doers from the dreamers. The game isn't yet over, but evidence is mounting that only the fittest will survive.

The basic choices of concept have been between horizontal and vertical axis configurations and between smaller and larger machines. Within these options, there are infinite choices relative to specific subsystems, components, parts and materials. Although most design efforts start with major attention to rotor blades and their aerodynamic performance, experience brings forth recognition that WECS are rotating electromechanical machines which must operate and survive in severe outdoor environmental conditions. A total system is far more than the assemblage of its physical parts.

The horizontal versus vertical axis consideration weighs advantages and disadvantages of both and should take into account the intended application. Maximum instantaneous conversion efficiency (power in the wind converted to rotor shaft power, usually referred to as power coefficient, Cp) for any theoretical WECS is generally accepted to be a little under 60 percent (the Lanchester–Betz limit) and an idealized horizontal axis wind turbine (HAWT) conceivably could reach that theoretical maximum at a very high rotor speed compared to windspeed. An idealized vertical axis wind turbine (VAWT) could approach a theoretical maximum conversion efficiency of around 55 percent. However, maximum expected aerodynamic conversion efficiencies of both HAWTs and VAWTs will be less than 50 percent in real world applications. Before testing of full scale machines, there was general belief that HAWTs were inherently more efficient. Field experience has shown that assumption to be false. VAWTs are every bit as efficient as HAWTs and, in some cases, more so. Figure 2-1 attempts to relate theoretical or idealized situations to actual test data reported in 1981 for DOE by NASA Lewis Research Center (HAWT) and Sandia National Laboratories (VAWT).

Aerodynamic conversion efficiency is a small part of the total picture in comparing one approach or machine to another. Ultimately, the key efficiency issue is useful energy generated per dollar invested. In that context, the two basic approaches offer both advantages and disadvantages.

VAWTs accept winds from all directions without the need for yawing to orient the rotor. They also utilize a natural aerodynamic stall to regulate power output without pitching the blades. By eliminating those two functions, VAWTs can be simpler in construction and can respond quicker to changes in wind direction or velocity, resulting in higher net efficiency of converting available winds to electricity. However, without blade pitching VAWTs need reliable brakes to stop and park the rotor in storm conditions.

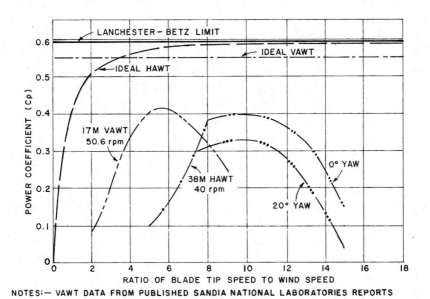

Figure 2-1. Aerodynamic performance of modern WECS.

Because both ends of VAWT blades are positively fixed to a rotating torque tube, as opposed to HAWT blades which cantilever from a pitching mechanism at a central rotating hub, VAWT performance in gusty or turbulent wind conditions is more steady and positive. VAWT blades and their attachments should be lower in cost and more rugged in operation. However, proper design and engineering are required with both to avoid excess fatigue stresses, and resulting deterioration over time, at blade joints and attachments.

Most VAWT working parts which may require maintenance or service are located at ground level, providing ease of access. Because the generator is also there, as opposed to the top of a tower, the electrical connections are easily made and electricity taken out directly at ground level. VAWT parts at the base are not limited by weight or bulk limitations which are inherent considerations in HAWT nacelles located and yawed on top of a tower.

The major advantage of HAWTs is that their rotors can be installed high in the unobstructed, less turbulent winds (where wind intensity is also usually higher than nearer the ground) by simply extending the tower height and reinforcing the supporting structure. VAWTs, with their guy wires and ground located working parts, cannot be as conveniently or economically raised. In addition, HAWT blades extend above the tower and attachment hub while the rotating tower (torque tube) of a VAWT is contained within the rotor to its top.

HAWT towers offer a choice between upwind and downwind rotor locations. The downwind position is the naturally stable configuration but introduces consideration of the wind shadow caused by the wind having to go through the tower before reaching the rotor blades. When blades are above the hub they see different winds than when they are behind the tower. Cyclic loads are accentuated. The upwind position choice eliminates the wind shadow problem but puts more demands on the yawing mechanism to hold the rotor in position. Gravity loads and cyclic stresses caused by the cantilevered blades rotating through varying wind loads between their high and low locations still must be addressed, but are more manageable. To deal with natural or resonant frequencies, design choices can be made between stiff and soft towers and rigid or teetering blade-to-hub connections.

Because VAWT towers are within the swept area of their rotors, lightning protection and hazard warning lights can be more easily and positively provided. VAWT guy wires, attached above the rotor, provide a natural path for lightning grounding and for wiring of the hazard lights. If HAWTs have more than two blades, or if one of two blades is parked above the nacelle, providing protection can be complex.

VAWTs define their swept areas with blades at the circumference and HAWTs define theirs with blades as the radius. Therefore, more blade material is usually required for VAWTs. VAWTs operate more efficiently at lower operating speeds than HAWTs, requiring higher step-up ratio gear boxes to drive high speed electrical generators. Positive electromechanical starting and braking must be provided with VAWTs, partially balancing the extra cost of HAWT yawing and pitching mechanisms.

Choice between smaller and larger machines in a WECS project should generally be made on the basis of the energy expected from specific available machines and the installed cost of the total project (as compared to just the cost of installing the machines). In some cases, space limitations will suggest larger machines if large quantities of energy are needed. In others, larger machines may be excluded because of visual or safety considerations. Availability of delivery and installation equipment may limit machine size at some sites.

The issue of large versus small often gets confused with the more basic questions of centralized versus dispersed locations of the WECS. Dispersed installations are more difficult for utilities to integrate into their systems and more difficult to operate, maintain, dispatch and, if necessary, to provide with storage. If dispersed WECS electricity has to be transmitted great distances, there will be line losses. If power lines aren't already in place, there will be extra cost. Development of a centralized WECS project should offer economies of scale and logistics and reduced redundant costs and time for site acquisition, engineering, and approvals.

Advocates of dispersed WECS installations (interconnected with the utility system through transmission and distribution lines) argue that machines can be placed nearer the energy utilizing loads, can be sited where the winds are best, and can take advantage of differing wind patterns (when the wind isn't blowing at one location, it may be at another) to provide more level power output. Dispersed installations also provide the opportunity for sharing costs and risks among multiple owners while keeping control and management of the total electrical system with the centralized utility. Windfarms by small power producers are an example of the latter advantage.

Whether central or dispersed, a WECS project can be large or small and can utilize large or small machines. Those choices should be made on the basis of maximum energy out per dollar in (AkWh/$), minimum investment for energy out ($/AkWh), best land use (AkWh/acre or AkWh/site cost) and availability and reliability of WECS hardware.

At one time, there was a general belief that bigger was better (more cost effective). However, examples supporting that conclusion have not yet been demonstrated.

If a project is not limited to one machine by power or energy maximums, the choice between fewer larger units or more smaller rotors is not always obvious. Larger machines have the advantage of fewer interconnections with the control or dispatching center and are considered by most utilities to be more compatible with existing generation, transmission and distribution methods. If the larger machines can also be purchased and installed at low cost, the total project can be less expensive.

Because of size and weight, larger machines are more demanding of delivery and installation equipment and people. Larger equipment is not always readily available or low in cost. Specialized worker skills are not always available locally. Importing of special equipment and people can be expensive. Without mass production of the larger machines, units in the early '80s were very expensive. WECS projects with large machines (few or many) generally have been dominated by on-site work and have been the equivalent of large construction projects. Cost effective installations are dependent on both the success of factory mass production and increases in on-site productivity expected to evolve with experience, continuity of operations and competent construction management.

Installation of many smaller machines (to provide the same energy capacity as fewer larger units) offers the economies of factory mass production, quality control, and prefabrication. Faster, less demanding, and lower cost on-site work should result. Some advocates believe that multiple smaller units provide insurance against downtime of a major portion of the project. The opportunity for exposure to many wind regimes might keep a higher portion of the power capability in service more of the time. However, advocates of larger machines counter with the argument that the more machines there are, the greater the opportunity for problems.

Regardless of machine size, multi-machine WECS allow incremental development of the total project. Individual machines can be brought on line quicker and start to generate revenue or savings faster. Negative cash flow can be minimized and better controlled. Risks can be minimized. The project can be modified or aborted in response to experience with early units. The first units can, in fact, be treated as research and development expenditures with attendant financial advantages.

Within the basic HAWT versus VAWT, centralized versus dispersed, large versus small, and few versus many choices, there are many technical questions relative to components, materials, type and quality of electricity (AC or DC, 60 Hz or 50 Hz, voltage, VARs, surge, etc.), and meteorological and structural criteria which can only be addressed by experienced professionals on a project-specific basis. Some of the more generic issues are also being addressed by government agencies and laboratories.

The major relative U.S. government support efforts are managed by the Departments of Energy (DOE), Agriculture (USDA) and Interior (DOI).

Basic technology development work is split among several subcontractors. The National Aeronautics and Space Administration (NASA) is DOE's technical arm for large HAWTs. Sandia National Laboratories (SNL) handles VAWT technology. Small WECS (less than 100 kW capacity) technology is managed at DOE's development and testing center at Rocky Flats, Colorado. Bonneville Power Administration (BPA) is serving as host utility for testing of DOE/NASA's three unit mini-wind farm (7.5 MW) near Goldendale, Washington. Pacific Northwest Laboratory (PNL) in Richland, Washington, is DOE's wind resource assessment program manager.

USDA concentrates on applications research at their Southwestern Great Plains Research Center in Bushland, Texas. DOI's major effort is with system verification of multi-megawatt scale WECS at their Bureau of Reclamation windfarm site near Medicine Bow, Wyoming.

If you want to get into the details of any of those programs, the appropriate agency should be contacted directly. Addresses and telephone numbers are included in Appendix C. Some U.S. technology milestones, as well as a scattering of foreign WECS accomplishments, are mentioned in Appendix A — WECS Historical Evolution.

TECHNICAL RISK AND IMMATURITY

The WECS industry can best be characterized as evolving from an embryonic 1970s base of technical risk and immaturity toward dynamic growth in the mid-'80s. Early efforts were dominated by technological experimentation and trial-and-error demonstrations. Government funded and guided research and development programs combined with liberal incentives led to overconfidence and premature commitments to hardware commercialization and project development. Pioneers and adventuresome entrepreneurs jumped in where wise men feared to tread. Few have been successful. Most have failed. For many, the outcome is not yet known.

Not surprisingly, the volatility of the emerging industry has caused dislocations, regroupings, and flexibility of response to constantly changing conditions. Some pioneering WECS producers have been acquired by more established firms. Promising technology has been sold by its inventors to others more capable of carrying innovation forward into commercialization. Scattered government programs have been consolidated into more responsible efforts to demonstrate effective solutions to known problems. Industry leaders have come and gone. Initial glamour and momentum have been replaced by economic realities and hard work.

Within the uncertainties which accompany immaturity, producers and project developers have had to decide both if and how they could survive during the shaking out and learning process. Without direct government or outside support, many dropped by the wayside. Some chose to let others lead while awaiting clarification or improvement of the financial, institutional, and political climates. Fortunately, a few chose (or perhaps couldn't find a better option) to plunge forward and develop their products, business, and projects while knowingly or unknowingly developing a WECS industry.

Most of the risk and maturity issues appear to fall into the general category of the chicken-and-egg dilemma: which comes first, mass production or mass purchases? How can production costs be low without volume? How can volume purchases be committed without low prices, proven performance, and meaningful warranties? How can performance be proven if full scale prototypes can't be built and tested over time?

In about twenty years, a few producers will be able to report on operating experience of about two decades, perhaps as much as 22 years. But modern commercially promising WECS hardware is so new that no current producer can claim more than a year or two of operating experience for his hardware. Naturally, potential buyers or investors want their WECS projects to last forever and perform reliably with minimum attention.

The pragmatic answer to the chicken-and-egg dilemma is, of course, a sharing of the risks by those who can ultimately take advantage of the rewards. What are the risks? What are the rewards? Who are the sharers?

By the early '80s, technical risks were well defined and relatively few in number. Proof-of-concept prototypes of many sizes and shapes of WECS were up (in some cases down again) and being tested in a wide variety of environmental conditions. Almost everyone involved in WECS technological innovation was aware that the only way to evaluate a WECS was to build a full scale machine and operate it under close scrutiny. Then the jobs of refining, simplifying, reducing costs, and improving performance could begin. Those who earlier harbored thoughts that they could go from design to a production model learned quickly not to fool with Mother Nature.

When WECS prototypes pass the test of survival and satisfactory short term generation of electricity, technical risk issues can be reduced to questions of lifetime, electrical stability and safety, operational interface and control, and survival in response to a wide variety of wind, weather, and site conditions. Will the WECS, in fact, produce usable electricity in reasonable quantities over a long period of time? At least until it pays for itself and provides an adequate return to its investors? The trick is to demonstrate in a relatively short period of time that complex hardware exposed to a sometimes hostile environment can operate in a reliable manner without causing problems to itself or its surroundings over a much longer period.

Beneficiaries of successful WECS obviously include WECS producers and owner-operators. Their ultimate rewards will primarily be in the form of profits or cost savings. All participants in the sale, delivery, installation, planning and execution of a WECS project, including suppliers of materials, labor, and services to the WECS producer, have the opportunity to share in those financial rewards. The public at large, as well as its government at all levels, will benefit if energy cost increases can be moderated and if the wind can be exploited as a practical alternative to imported, depletable, or potentially polluting or hazardous fuels. The societal, economic, and national security advantages of harnessing the locally available, free, renewable winds are self evident.

So who should share in the technical risk of proving promising new technology in the innovative WECS industry? Just about everybody. And they have been sharing, whether they knew it or not.

Both the U.S. and Canadian governments have sponsored direct WECS research, development and demonstration projects which have helped get the producers and users over the first hurdles. The U.S. government and many states have provided tax credits for investments in WECS which help reduce financial risks. Some states have also used public funds to provide subsidized loans and other incentives, as well as direct purchases of untested hardware, to provide experience, to build confidence and help get momentum going toward widespread utilization.

In the private sector, the American Wind Energy Association (basically representing direct WECS participants) and a variety of industry and professional associations have worked toward establishing product and performance standards and disseminating information to assist both producers and potential users.

The three major electric utility groups (investor owned, publicly owned, and rural electric cooperatives) have sponsored WECS demonstration projects and have actively monitored progress and problems.

The Electric Power Research Institute (EPRI) has concentrated on monitoring progress of WECS hardware and prototype installations so that experience with a few machines could be of value to the entire utility industry. EPRI has been mainly concerned with WECS performance and reliability and the impact of WECS integrated into utility generating systems or interconnected with utility transmission and distribution systems. Although experience has been limited, EPRI has not found major problems or reasons why WECS cannot successfully complement the nation's electric utility system.

The American Public Power Association (APPA), which counts as members most of the federal, state, and municipal utilities as well as a few rural electric cooperatives, took an early leadership role by funding a hands-on WECS demonstration project through their research committee in early 1979. With broadbased cooperation involving many of the municipal utilities, people's utility districts, and rural electrical cooperatives in Oregon, APPA's initiative led to installation in late 1980 of the world's largest VAWT on the windy Oregon coast, above Agate Beach. Continuing research, retrofitting, and testing of that Alcoa-built 500 kW prototype demonstrated the feasibility of locating WECS at windy sites and moving the electricity through existing utility power lines to customers who can use it. By involving a large group of participants, the technical risks and cost of innovation were relatively small to each, and the evolving base of knowledge and confidence can be quickly exploited when the system proves itself.

The National Rural Electric Cooperative Association (NRECA) combined the monitoring and information dissemination roles of EPRI and the hands-on experience role of APPA. Because most RECs are not electricity generators, NRECA's research efforts have been directed more at dispersed installations of smaller WECS at members' sites. Particular emphasis has been on evolving standards for safe interconnection of WECS to distribution lines and insurance ramifications of the split responsibilities of cooperative members, the cooperative, and the supplying utility.

In addition to monitoring and reporting on numerous WECS installations, NRECA's research and development committee joined in sponsoring an innovative dispersed WECS test program with the Pennsylvania Rural Electric Cooperatives Association in 1979. Allegheny Electric Cooperative, Inc., as the supply utility for 14 Pennsylvania and New Jersey cooperatives, became project manager and, ultimately, owner of the WECS hardware. The costs were equitably shared by all of the participants. Two small WECS test installations were planned. An 8 kW VAWT prototype was intertied with a three-phase distribution line in western Pennsylvania at a site served by Southwest Central Rural Electric Cooperative Corp. (See Figure 2-2.) The second unit was intended to test a WECS tied to a single-phase distribution line in the mountains near Gettysburg at an Adams Electric Cooperative site. NRECA provided instrumentation and funds for data retrieval and reporting. Unfortunately, the Gettysburg area

Figure 2-2. NRECA Pennsylvania VAWT research project.

project was not installed because of lack of single-phase WECS hardware. And the 8 kW VAWT prototype was removed from the Southwest Central site in late 1981 because of hardware problems and lack of sufficient winds in western Pennsylvania to make further investment in research attractive. In this case, the technical risk was reduced by cost sharing by three RECs, NRECA, and the hardware supplier. However, poor siting and unsuccessful hardware combined to abort the ambitious project prematurely.

In addition to the government and industry efforts, a few pioneering utilities have installed larger WECS to make their own assessments of technology and its effect on their systems. Some of the specific projects are described in more detail in the utilities section (Chapter 3 – Bulk Power) of this book. Utility contributions to developing and proving new WECS technology and overcoming traditional fear of innovation are truly remarkable. In the public sector, both DOI's Bureau of Reclamation (6.5 MW at Medicine Bow, Wyoming) and the Bonneville Power Administration (7.5 MW at Goldendale, Washington, plus many small units) have pioneered large-scale windfarms integrated with major hydroelectric systems. In the private sector, Southern California Edison (3.6 MW in three of their own research prototypes and cooperation with small power producers in numerous windfarms), Pacific Gas & Electric (2.5 MW in their own research machine and cooperation with small power producers and the California Department of Water Resources in both windfarms and additional research projects), Pacific Power and Light (200 kW research machine with more to come) and approximately 100 other utilities (measuring their wind resources, analyzing WECS feasibility or cooperating with private project developers) have invested R&D funds to help establish WECS as viable additions to their total electrical systems. The private utility experience is especially valuable because of the general distrust of government generated data in areas of innovative technologies.

In addition to helping reduce technical risk, procurement and installation of WECS at effective locations around the country have had significant impact on potential producers, systems providers and system and energy users. The extent of that impact has been highly dependent on the choice of system hardware, site conditions, and the method of integration with existing power systems.

As those pioneering projects accumulate operating time, and as additional local installations evolve, growth toward a mature WECS industry is currently taking place. The innovative proof-of-concept efforts serve to:

- Demonstrate satisfactory operational history and allow establishment of standards for safety, reliability, performance, and durability.
- Demonstrate credible procurement, operating, and maintenance economics and provide experience to local engineering, installing, and operating personnel.

- Demonstrate electricity generation as a function of time of day, time of year and geographic location, to permit more dependable planning of output and integration with utility operations.
- Make electricity or WECS hardware available to utilities, or direct to consumers, at low risk with minimum development or acquisition cost.
- Provide the basis for long-range planning for WECS/utility integration and procurement decisions for hardware and land purchases, and other long lead time investment decisions.
- Result in identification of institutional constraints which need modification to permit development of locally beneficial WECS capabilities.

As initial hardware and applications prove themselves, both producers and users are gaining confidence and moving toward larger scale and commercially oriented WECS projects in high-wind, high-fuel cost areas. These installations represent an early market to build up the industry infrastructure as well as user acceptance. As the perceived risk is reduced, more advanced systems will be developed and production capabilities will expand. Confidence, increased demand, hardware refinements, and local experience with WECS will help hold the line on system costs and make WECS more cost effective in broader markets as fuels competitive with wind rise in price.

RESEARCH AND DEVELOPMENT THRUST

The research and development thrust has evolved from the glamorous to the practical, and more toward development than research. By the early 1980s, WECS had been demonstrated in sizes ranging from less than 1 kW to over 4 MW. Both upwind and downwind HAWTs, with almost any number of rotor blades from one to six, demonstrated they could work. In VAWTs, both straight bladed and Darrieus type (curved troposkein shaped blades) rotors, with both two and three blade configurations, had been operated and tested. A variety of approaches to key components (such as blades, speed increasers, generators, controls, etc.) had been tried. The question of WECS generating useful electricity was answered. They could.

So the R&D thrust for the '80s evolved to developing more cost-effective, reliable, and safe WECS and demonstrating those characteristics in real world applications. Experience proved that until a WECS design was built and tested, it was without credibility. No matter how good the design team or the technological tools, little correlation was demonstrated between paper designs and installed hardware.

The '70s provided a long laundry list of WECS technology from which to choose a starting point for commercial hardware.

U.S. Government efforts at DOE's Rocky Flats small WECS test site provided a public domain data base for WECS designs ranging from very small residential machines through 8 kW, 15 kW, 40 kW, and 100 kW rated systems for agricultural and commercial applications. DOE's Sandia National Laboratories successfully demonstrated Darrieus type VAWTs as small as 2 kW and as large as 100 kW and put those prototypes into test programs. (See Figure 2-3.) DOE's

Figure 2-3. Sandia National Laboratories Darrieus VAWT Research Center.

large HAWT research efforts, through NASA, resulted in research machines of 100 kW, 200 kW, 2 MW and 2.5 MW capacity which are being tested and retrofitted as needed.

Other government efforts have resulted in major working prototypes in Australia, Denmark, Canada, England, France, Germany, Holland, Ireland, Italy, Russia, Spain, and Sweden. Appendix A (WECS Historical Evolution) describes some of those prototypes. Although foreign design requirements, particularly electrical, are not exactly the same as in the United States, much has been learned from international experience and the reported results of those experiments.

Outside the government efforts, pioneering private sector research resulted in valuable prototypes by U.S. companies such as Wind Power Products Company (Charlie Schachle's innovative 140 kW HAWT in Moses Lake, Washington, 1977), Grumman (their Windstream 25 served as the basis for numerous R&D projects in the early '70s), Energy Development Company (the late Terry Mehrkam's vehicle for prototyping numerous small and large HAWTs in a variety of applications), Jay Carter Enterprises (development of their 25 kW HAWT in the late '70s, now being commercialized under the name of Carter Wind Systems), WTG Energy Systems (their 200 kW HAWT prototype on Cuttyhunk Island, Massachusetts in 1977 provided encouragement that larger machines could, in fact, be designed and prototyped at reasonable costs without government help) and Alcoa (their leadership in funding practical development efforts, based on U.S. and Canadian research, culminated in installation of three versions of the world's largest VAWT in the early '80s).

As many North American companies now move toward commercialization of WECS hardware, the roles of Canadian and U.S. governments in WECS R&D has become more focused. In the U.S., the Reagan administration announced that it planned "to restructure the technology programs of the Department of Energy to emphasize longer-term, high-risk, potentially high-payoff R&D, while terminating large technical demonstrations."

A part of the administration's policy is to use tax incentives and the decontrol of oil and gas, which will result in higher fuel prices, to motivate private corporations to pick up the R&D projects no longer funded by the federal government, particularly those projects in the demonstration phase. The DOE policy was summed up by

Secretary Edwards when he stated, "I think government's role is to develop the technology and prove that it works and then the commercialization should be in the private sector."

At the implementation level, Dr. Louis V. Divone, the federal government's longtime wind program manager, acknowledged in his presentation at DOE's 1981 fifth biennial wind workshop that a significant change is occurring in the role the federal government plays in the development of wind energy. However, Divone pointed out that "It is still the sunrise not the sunset of windpower in America. While the federal wind energy program's objective remains the same — to enable wind energy to be used on a significant scale — its efforts have been redirected toward technological research in high-risk, potentially high-payoff areas, and towards research beneficial to all manufacturers and users."

Dr. Divone further elaborated by stating, "The new thrust of the program is to provide the technology and information base needed by industry. Research will continue in key areas, including aerodynamics, structural dynamics and fatigue phenomena, reliability and multiple systems interactions."

Most private sector WECS producers, and potential producers, have welcomed the new government role. Wind energy has so much potential that a small amount of government support over the next few years should yield a vibrant and self-sufficient industry.

That "small amount of government support" is still very important during the transition from the R&D phase into an embryonic commercialization stage. DOE is still needed to gather and report wind resource data. DOE is still needed to test and report results of the Rocky Flats, Sandia, and NASA prototype testing. And DOE is still needed to remove the institutional barriers and to serve as the advocate within government for appropriate tax credits, utility regulations and incentives, and government purchases of WECS.

Within the category of "longer-term, high-risk, potentially high-payoff R&D," the future of DOE's efforts and, in fact, the type of government sponsored R&D needed by the producers, is not entirely clear. Because of long lead times dictated by government planning, budgeting, authorizing and executing functions, DOE is still saddled with old programs, obsolete prototypes, and commitments to funding of currently questionable projects. At this writing, the DOE WECS

R&D program for fiscal year 1983 is still being argued. Over five million dollars is expected to be budgeted, and the challenge is to allocate funds in a way that will provide the most leverage for the WECS industry.

In the meantime, major U.S. government WECS R&D efforts are continuing at Sandia and NASA.

After having its attempt to move to larger VAWTs (the MOD-6V) scuttled by budget cuts in 1981, Sandia has concentrated on its first (approximately 60 kW capacity) and second (approximately 100 kW capacity) generation 17 metre diameter VAWTs. (See Figures 2-4 and 2-5.) Sandia is attempting to solve problems uncovered in its

Figure 2-4. Sandia 17 metre VAWT prototype.

Figure 2-5. Sandia low cost 17 metre VAWT.

four prototypes (Albuquerque, Rocky Flats, Bushland, and Martha's Vineyard) and to make those machines more productive, reliable, and safe while increasing life expectancy. In addition, it has continued its impressive program of information dissemination and VAWT technology transfer with industry. At this writing, Sandia has active technology transfer agreements with Alcoa (Pittsburgh, PA), Flow Industries (Kent, WA) and Forecast Industries (Albuquerque, NM) who are attempting to develop and commercialize their versions of the DOE/Sandia VAWTs.

NASA is continuing its efforts to grow its HAWT technology to gigantic machines which they maintain have the potential for generating electricity at competitive costs if machines are constructed in volume. After mixed results with their smaller (100 kW MOD-0, 200 kW MOD-0A, 2 MW MOD-1 and 2.5 MW MOD-2) research machines, (see figures 2-6 and 2-7) NASA is now concentrating R&D efforts on multi-megawatt capacity MOD-5s. Two versions of the MOD-5 are being detailed and analyzed. MOD-5A is being designed for NASA by General Electric and is currently sized to generate approximately 6.2 MW at 20.4 mph. MOD-5B is being designed for NASA by Boeing and is currently sized to generate approximately 7.2 MW at 20.5 mph. It is unclear whether either, or both, of those designs will be prototyped with government funds. Whether they are or not, the evolving public domain technology should be useful to many producers who are pursuing larger HAWT hardware.

In Canada, the R&D is even more focused. The National Research Council (NRC) has concentrated on Darrieus type vertical axis wind turbines since 1966. With 50 kW and 500 kW capacity VAWTs now being offered commercially (see information on DAF-Indal's products in the "Producers" section of this chapter), NRC is now in the process of developing the world's largest VAWT. Known as Project Aeolus, the $35 million development project involves one of Canada's leading utilities (Hydro-Quebec) and is expected to result in a prototype machine with a strut-reinforced, two-blade rotor approximately 350 ft high with a generating capacity in the 3–4 MW range. That size machine was chosen as representative of multi-megawatt scale machines up to 8 MW in capacity. Earlier detailed studies for NRC by Shawinigan Engineering and Canadaire Ltd. projected lowest cost energy from machines in that size range.

Regardless of who funds or does the research, the objectives are now clear. WECS must prove themselves to be capable of:

- Producing quality electricity in reasonable quantities.
- Operating in a safe and reliable manner.
- Operating without causing utility system interface problems.
- Surviving for 20–30 years without excessive operating and maintenance costs.
- Installing at total costs low enough to be widely affordable.

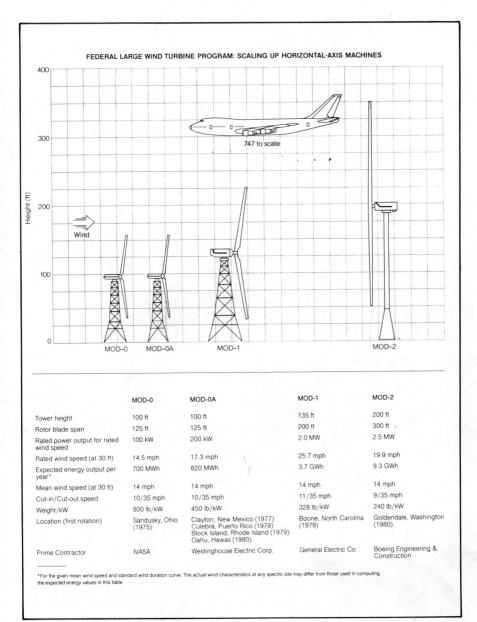

FEDERAL LARGE WIND TURBINE PROGRAM: SCALING UP HORIZONTAL-AXIS MACHINES

	MOD-0	MOD-0A	MOD-1	MOD-2
Tower height	100 ft	100 ft	135 ft	200 ft
Rotor blade span	125 ft	125 ft	200 ft	300 ft
Rated power output for rated wind speed	100 kW	200 kW	2.0 MW	2.5 MW
Rated wind speed (at 30 ft)	14.5 mph	17.3 mph	25.7 mph	19.9 mph
Expected energy output per year*	700 MWh	820 MWh	3.7 GWh	9.3 GWh
Mean wind speed (at 30 ft)	14 mph	14 mph	14 mph	14 mph
Cut-in/Cut-out speed	10/35 mph	10/35 mph	11/35 mph	9/35 mph
Weight/kW	800 lb/kW	450 lb/kW	328 lb/kW	240 lb/kW
Location (first rotation)	Sandusky, Ohio (1975)	Clayton, New Mexico (1977) Culebra, Puerto Rico (1978) Block Island, Rhode Island (1979) Oahu, Hawaii (1980)	Boone, North Carolina (1979)	Goldendale, Washington (1980)
Prime Contractor	NASA	Westinghouse Electric Corp.	General Electric Co.	Boeing Engineering & Construction

*For the given mean wind speed and standard wind duration curve. The actual wind characteristics at any specific site may differ from those used in computing the expected energy values in this table.

chart courtesy EPRI Journal

Figure 2-6. NASA HAWT research.

Figure 2-7. NASA test bed research HAWT.

Not only must successful research include installation and testing of full scale prototypes, that testing, as well as successful performance, must take place at sites with good winds in typical "commercial" conditions before most utility planners will conclude that the technology is ready for widespread utilization in their systems. The proof of research and development efforts can only occur when machines are installed, debugged, operated under controlled conditions, refined,

and then operated in automatic and unattended, but carefully monitored, modes for at least two or three years. Only then can responsible
conclusions be made about the success of new WECS technology.

PRODUCERS OF ELECTRICITY MACHINES

Fortunately, relative WECS R&D started in the '60s and has been continuous since. As a result, a few producers of electricity machines are
ready, or getting ready, to fabricate and deliver their WECS hardware
for installation in commercial and/or demonstration projects. The remainder of this chapter discusses the known U.S. and Canadian firms
believed committed to commercialization of WECS of substance (arbitrarily defined as those with electricity generating capacities of 20 kW
or greater).

The emerging commercial WECS hardware is grouped into three
categories — small, intermediate, and large. The very small (less than
20 kW capacity) are covered in detail in other books and many government reports and listings.

In this book, small is defined as 20–75 kW. The current producers
include Carter Wind Systems, Energy Sciences, Enertech, and Wind
Power Systems. Additional probable producers, either in the start-up
or development stage include PM Wind Power, WECS-Tech, Wind
Engineering, and Windtech. In the wings, with technology ready but
without announced commercialization commitment, are Butler
Manufacturing Co. (40 kW HAWT) and Kaman Aerospace (65 kW
HAWT). In addition, Fayette Manufacturing (50 kW HAWT) and
U.S. Windpower, Inc. (50 kW HAWT) produce machines in this size
range for use in their own windfarms and DAF-Indal has been providing a 50 kW VAWT for research projects. (See Figures 2-8 through
2-11.)

Intermediate size WECS are defined as those with generating capacities in the 76–750 kW range. Current producers include DAF-Indal,
Westinghouse Electric, and WTG Energy Systems. Producers in the
start-up or development stage include California Energy Group, Flow
Industries and Forecast Industries. Alcoa has its 500 kW VAWT on
hold. Rockwell International and McDonnell Douglas both reportedly have intermediate size WECS on the drawing boards but neither
has made a commercial offering. At least two European WECS (the

Figure 2-8. Butler Manufacturing 40 kW HAWT prototype based on Mehrkam Energy Development Company technology.

Volund 265 kW HAWT from Denmark and the 50, 100, and 150 kW HMZ HAWTs from Belgium) are reportedly being considered for projects in North America.

Large WECS, those over 750 kW in capacity, are expected to be available soon from Hamilton Standard Division of United Technologies. Bendix Corporation is reportedly working on a multi-megawatt scale

Figure 2-9. Kaman Aerospace 65 kW HAWT prototype.

HAWT in their Aerospace Division but has abandoned the 3 MW design which was prototyped with Southern California Edison. Both Boeing and General Electric are continuing R&D work with the DOE/ NASA MOD-5 program and Boeing is continuing to monitor and refine the DOE/NASA MOD-2 design and their five prototypes.

The Producers of Small WECS (20 kW to 75 kW in Capacity)

The four mid-1982 producers of small WECS have all faced the test of real world exposure to their products and each has gone through

Figure 2-10. U.S. Windpower 50 kW HAWT prototype.

the trials and tribulations of servicing and retrofitting their hardware before getting it right. After many setbacks, they appear to be surviving and their hardware is being offered for commercial use.

Although wind machines in this size range were originally thought to be most useful in agribusiness and irrigation applications, most of the early shipments have been multiple units to small power producers (SPP) for use in windfarms.

The four small WECS industry leaders are Carter Wind Systems, Energy Sciences, Enertech, and Wind Power Systems. (See Table 2-1.)

Figure 2-11. DAF-Indal's 50 kW research VAWT.

Carter Wind Systems. Formerly known as Jay Carter Enterprises, this Burkburnett, Texas company started working on their 25 kW two-blade downwind HAWT in 1975, emphasizing simplicity, ease of erection and maintenance, their inhouse expertise with glass fibre reinforced plastics, and established induction motor-generator technology.

The Carters (Jay, Sr. and Jay, Jr.) developed their machine and prototyped and tested it over four years in close cooperation with the U.S. Department of Agriculture's (USDA) wind research center in Bushland, Texas and with the Alternative Energy Institute at West Texas State University.

Table 2-1. Current and Expected Producers of Small WECS (20 kW to 75 kW).

WECS PRODUCER	POWER (kW)		ROTOR			
	CAPACITY	RATED (WINDSPEED)	SIZE	SPEED	BLADES	TYPE
Carter Wind Systems	30	25 (26 mph)	32 ft diam.	120 rpm	2	downwind HAWT
Energy Sciences	65	50 (30 mph)	54 ft diam.	76 rpm	2	downwind HAWT
Enertech	24	20 (21 mph)	44 ft diam.	53 rpm	3	downwind HAWT
Wind Power Systems	40	27 (25 mph)	39 ft diam.	129 rpm	3	downwind HAWT
PM Wind Power	30	25 (25 mph)	34 ft diam.	120 rpm	2	downwind HAWT
WECS-Tech	75	70 (28 mph)	49 ft diam.	68 rpm	3	downwind HAWT
Wind Engineering	25	25 (25 mph)	42 ft diam.	68 rpm	3	downwind HAWT
Windtech	20	20 (35 mph)	32 ft diam.	—	2	downwind HAWT
Windtech	40	40 (35 mph)	48 ft diam.	—	2	downwind HAWT
DAF-Indal	55	50 (40 mph)	55 ft × 37 ft	80 rpm	2	Darrieus VAWT

The development process was not smooth. There were several machine and component failures and frequent retrofits of hardware in service. But the Carters stayed with it and built a reputation for reliable service and a low cost product. They built their business to a point where, by early 1982, they had a backlog of over a thousand units sold and were gearing up for production of thirty units per month.

Carter's Model 25 is rated 25 kW at approximately 26 mph wind velocity. Maximum power is 30 kW in approximately 30–40 mph winds. The two-blade rotor sweeps a 32 ft diameter and drives either a single-phase or three-phase induction generator. Hub height is 60 ft. It is designed for erection by a two-person crew without the need for a crane. Unlike most other designs, the Carter 25 continues to operate and generate electricity in storm winds. In high winds, the blades flex and regulate torque. Brakes are only expected to activate in grid loss or other emergency conditions. (See Figure 2–12.)

Carter sales success with Model 25 has encouraged them to consider intermediate size machines and they have explored both 125 kW and 500 kW models. Their sales success also attracted a capital infusion by Hamilton Standard, a division of giant United Technologies Corporation. Although details of the Hamilton Standard/Carter agreement were not announced, the new capital should allow Carter to proceed with development and testing of its Model 125.

Carter Wind Systems Model 125 will essentially be a scaled-up version of Model 25. The intermediate size HAWT is expected to be rated 125 kW at 30.5 mph wind velocity. Its maximum power is expected to be 150 kW in 40 mph winds, measured at its 120 ft hub height.

Model 125 will have most of the features of Model 25 but its rotor diameter will be increased to 64 ft and will operate at 75 rpm. Only three-phase AC current will be available from the induction generator.

Energy Sciences, Inc. ESI was founded in 1980 by individuals who got their basic WECS training as employees of DOE's small wind energy conversion systems (SWECS) test center at Rocky Flats, Colorado. The group, headed by Jim and Sharon Alexander, Sandy Butterfield, and Jim Sexton located their new business in nearby Boulder, Colorado and made a deal with CL Industries for fabrication of key components by CL's Wintec subsidiary.

Figure 2-12. Carter Model 25 HAWT.

ESI designed a 50 kW (at 30 mph) HAWT by selecting the best features of public domain technology which was evolving from both DOE's large HAWT R&D program (administered by NASA) and the small HAWTs evolving from DOE's Rocky Flats R&D. By taking advantage of experience by others, they were able to cut years from the normal WECS development process.

The ESI-54 is a 54 ft diameter, two-blade, downwind HAWT. It utilizes wooden blades which rotate at 76 rpm to drive a 75 hp three-phase induction generator. Hub height can be either 60 ft or 80 ft, depending on choice of Rohn tower. All components were chosen as readily available off-the-shelf items for low cost and ease of maintenance and repair. Like the Carter 25, the ESI-54 is designed to be site erected without the need for a crane.

Although the 50 kW model was originally chosen because of its suitability for irrigation and other agricultural applications, its initial sales were to windfarmers in multi-unit quantities for installation of clusters. As a result, ESI directed special attention to control strategies and power factor correction to make their wind machines perform efficiently in such multi-unit situations.

ESI's first prototype was installed on Molokai (Hawaii) in early 1981 as the first of a planned five-unit windfarm by Carl Huntsinger's Molokai Energy Company. Their first windfarm prototype installation was in 1982 with ten units installed by Larry Larsen's Pacific Wind and Solar in the Tehachipi Mountains as part of the Ridgeline windfarm for Southern California Edison Company. (See Figure 2-13.)

Test results from the Molokai prototype, installed on a 60 ft tower, were very encouraging in both electrical performance and in resistance to the Pacific salt spray exposure. The 50 kW output at 30 mph (measured at 30 ft) was confirmed and power as high as 65 kW was observed in 45 mph winds.

Loss of one of the Tehachipi units was attributed to a bad brake and the unit was rebuilt. At the same time, an additional prototype was being tested at a higher rpm (approximately 90) to determine if the HAWT could be uprated to 70–75 kW in the future.

By mid-1982, ESI was booking volume orders for delivery of their 50 kW ESI-54 for several different windfarms.

Enertech Corporation. Unlike Carter and ESI, which started out with technology and the design of specific WECS hardware, Enertech started as a distributor of very small HAWTs produced by others, mostly in foreign countries. Because of their market orientation, Enertech today enjoys the industry's most complete dealer network with capabilities in sales, service and maintenance.

Enertech was founded in 1975 by Bob and Jane Sherwin, Bill Drake, and Ned Coffin in Norwich, Vermont. After experience distributing other companies' hardware, they developed their own 1.5 kW (The Enertech 1500) HAWT in 1977 and pioneered the concept

Figure 2-13. Energy Sciences 50 kW HAWT.

of utility connected induction generators. By 1981, the 1500 had been uprated to 1.8 kW (Enertech 1800). Most technical problems had been debugged and production capacity had grown to 200 units per month. They later added a 4 kW model.

With a unique capability to be caustically critical of the U.S. government's wind programs while using DOE's funds for their own research, development, and demonstration, Enertech developed a high-reliability 2 kW HAWT for remote stand-alone applications and a 15 kW HAWT intended for farms and light industry. The 15 kW contract with DOE's Rocky Flats SWECS organization (worth approximately $800 thousand) led to today's 20 kW HAWT, the Enertech 44.

Enertech's rapid growth and market leadership led to a need for outside capital to sustain the growth. Their success with the very small HAWTs, along with the potential for the evolving 20 kW machine, attracted a $750 thousand loan from Bendix Corporation which was later converted to a 30% equity position. Bendix has provided additional funds to sustain Enertech's growth in the small WECS while concentrating its own aerospace group efforts on multimegawatt scale HAWTs.

The Enertech 44 is unique in that it appears to be the first WECS designed with DOE funds to become a commercial reality. Work started in mid-1979 with the goal of developing a "reliable low-cost wind machine suitable for direct utility interface and capable of annual output of at least 50 thousand kWh in a 12 mph wind regime" for DOE. A prototype was tested and then delivered to Rocky Flats in late 1981. Additional pre-production prototypes were built and tested in early 1982 with full production scheduled to begin in mid 1982.

The 20 kW (at 21 mph) HAWT has three downwind laminated wood blades which sweep a 44 ft diameter. The fixed pitch blades rotate at 53 rpm to drive either a single-phase or three-phase induction generator. Hub height is at 80 ft with a self-supporting truss-type tower. Maximum power is 24 kW in approximately 30 mph winds. (See Figure 2-14.)

With their established dealer network and Bendix's financial backing, Enertech is in a good position to quickly become a leader in the small WECS industry, in addition to their established leadership role with the very small machines.

Figure 2-14. Enertech 20 kW HAWT.

Wind Power Systems, Inc. Founded in San Diego in 1974 by Ed Salter to design cost-effective and reliable WECS, Wind Power Systems has been selling their machines under the Storm Master brand.

Although WPS did not receive direct government funds for its R&D efforts, its machines were selected as part of DOE's ill-fated field evaluation program. In 1980 and 1981, ten small HAWTs were installed as part of that program, which was intended by DOE to provide utilities and users with experience with a wide variety of hardware in many different site conditions. Like most other WECS of that era, the Storm Masters had their share of problems requiring service and retrofitting. In addition to the ten units sold to DOE, additional units were shipped to private California and Hawaii sites.

By mid-1981, WPS had responded to feedback from the installed prototypes and was able to refine its design into a production model and start to take contingent commercial orders.

The new design is identified by WPS as Storm Master Model 12 Wind Power Plant. It has three downwind glass fibre/foam composite blades which sweep an area with a 12 metre (39.3 ft) diameter. The rotor operates at 129 rpm to drive a three-phase 45 kVA capacity induction generator. The HAWT is rated 27 kW at approximately 25 mph but its power continues to rise with increasing wind velocity until it is braked to a stop. The WPS power curve shows 40 kW at 60 mph.

Although Wind Power Systems did not make a smooth or quick evolution into commercialization of its Storm Masters, it was able to survive the shake-out of the early '80s. In 1982 it was the supplier of WECS hardware for the first phase of the Zond Systems, Inc. wind park for Southern California Edison in Tehachapi Pass, California. (See Figure 2-15.)

Potential Small WECS Producers. In addition to the four current producers (Carter, ESI, Enertech and WPS) at least four companies are working on designs or prototypes which could soon lead to commercial offerings of small WECS.

PM Wind Power Inc. is the energy subsidiary of Production Machinery Corporation in Mentor, Ohio. They have prototyped and are testing a two-blade downwind HAWT. (See Figure 2-16.) The glass fibre reinforced plastic blades operate at 120 rpm and sweep a 34 ft diam-

Figure 2-15. Wind Power Systems 27/40 kW HAWT.

eter circle to drive a single-phase or three-phase induction generator. The HAWT is rated 25 kW at 24.5 mph wind velocity. Maximum power is 30 kW in 27–36 mph winds. A flexible hub strap provides progressive stall of the airfoil and increased rotor cone angle as wind-speed increases. No specific shutdown windspeed is planned. The PM prototype is mounted on a 60 ft high galvanized steel monopole. With four guy wires and a hinged tower, the assembled HAWT can be raised and lowered utilizing a gin pole. No crane is required and maintenance can be performed at ground level. The producer projects that 40 thousand kWh per year can be generated in a 12 mph

Figure 2-16. PM Wind Power's 25 kW HAWT prototype.

regime and that a 15 mph site would yield approximately 55 thousand annual kWh.

WECS-Tech Corporation has a production out facility in Compton, and marketing offices in Lawndale, California. It is pursuing a HAWT design which utilizes the dacron covered sail wing concept earlier developed at Princeton University under the leadership of Dr. T.A. Sweeney. The WECS-Tech Model 7495 is described as a three-bladed, downwind, free yaw sail wing design with speed control tip-spoilers. The initial design is rated 70 kW at 28 mph wind velocity. The rotor has a diameter of 49 ft and operates at 67.8 rpm to drive a three-phase 100 hp induction generator. (See Figure 2-17.) The tip-spoilers cut in at 32 mph and drum brakes are activated at 40 mph to shut the rotor down in storms. WECS-Tech has a prototype in Compton and reportedly is proceeding with as many as three additional prototypes in Texas. WECS-Tech is preparing to develop a 5.5 MW wind park for Southern California Edison in the San Gorgonio Pass near Palm Springs when their hardware is proven.

Wind Engineering Corporation has been working since 1977 to develop, test, debug, and get its Wingen 25 ready for production. Founder Coy Harris has had more than his share of misfortune and setbacks, but believes production can begin in 1982. Located in Lubbock, Texas, Wind Engineering has worked closely with the

Figure 2-17. WECS-Tech 70/100 kW HAWT. Prototype is 70 kW, production model is projected as 100 kW.

USDA WECS research center in Bushland and West Texas State University's Alternative Energy Institute in developing its technology and testing its prototypes. Wind Engineering lost the rotor from its first prototype and was then delayed by lack of capital and extended acquisition negotiations with Bendix Corporation (which were later terminated without positive action). With new venture capital into the business in 1981, the Wingen 25 came back to life with a new and improved prototype and plans for a production line which could fabricate 20 to 30 machines per month. The new unit, Model 25–42, has been designed to either generate electricity or supply rotary mechanical power at ground level. The three-bladed downwind rotor has a diameter of 42.5 ft which operates at 68 rpm to drive either a single-phase or three-phase 30 hp induction generator. The HAWT can be mounted on 60 ft, 80 ft, or 100 ft high freestanding Rohn towers. The rated and maximum power is 25 kW which is achieved at wind velocities of 25–45 mph. The variable-pitch fabric-covered aluminum blades are feathered at 45 mph for safety during storms. (See Figure 2-18.)

Windtech, Inc. is a start-up company in Glastonbury, Connecticut. It was started in 1982 by Kip Cheney who left the United Technologies Research Center (UTRC) to pursue scaled-up versions of the bearingless rotor HAWTs developed by UTRC for DOE as part of the Rocky Flats SWECS program. Cheney had been UTRC's Project Manager and champion for their small HAWT R&D and prototyping effort. The two public domain UTRC/DOE HAWTs were rated 8 kW and 15 kW by DOE, both at 20 mph wind velocities. Cheney plans to uprate Windtech's versions to 20 kW and 40 kW by increasing the operating rpm and utilizing larger generators on each. The smaller machine will also have 1 ft added to the rotor diameter. Windtech, therefore, is pursuing two-bladed downwind HAWTs with a rotor diameter of 32 ft to drive a 25 hp induction generator (20 kW model) and a rotor diameter of 48 ft to drive a 50 hp induction generator (40 kW model). Most other details will be similar to the UTRC units prototyped for DOE. (See Figure 2-19.)

DAF-Indal, Ltd. has provided 50 kW VAWTs for two California research projects as well as for Canadian and foreign WECS projects. However, that Mississauga, Ontario firm is concentrating its commercial efforts on an intermediate size (500 kW) VAWT as described in

Figure 2-18. Wind Engineering 25 kW HAWT.

Figure 2-19. Windtech/UTRC prototype. Windtech 20 kW and 40 kW models are based on UTRC 8 kW and 15 kW research prototypes for DOE.

the next section. The 50 kW research machine is a two-bladed Darrieus type VAWT with a unique bull gear and hydraulic braking system. (See Figure 2-20.) It evolved from work started in 1974 and from experience with smaller and less sophisticated earlier models, such as used by USDA in their applications research work. Two of the newest models are operating in California, one at the San Luis reservoir as part of a California Department of Water Resources and Pacific Gas and Electric Company demonstration (installed in early 1981)

Figure 2-20. DAF-Indal "power module," containing innovative bullgear and hydraulic braking system.

and one at Southern California Edison's wind research site in the San Gorgonio Pass near Palm Springs (installed in early 1982). The Darrieus rotor is 55 ft high with a diameter of 37 ft. It operates at 80 rpm to drive either a 1200 rpm 60 Hz or a 1000 rpm 50 Hz three-phase induction generator. (See Figure 2-21.) As a further innovation, DAF modified one of their units (slowed it to 71.4 rpm and removed the generator) and integrated it with two 46 kW diesel generators for a prototype 100 kW "wind turbine assisted diesel generator system" which was installed near Sudbury, Ontario in 1982. (See Figure 2-22.)

Figure 2-21. DAF-Indal 50 kW VAWT.

The Producers of Intermediate Size WECS
(76 kW to 750 kW in Capacity)

The three producers offering commercial WECS in the intermediate size range in 1982 are DAF-Indal, Westinghouse Electric and WTG Energy Systems. Another three (California Energy Group, Flow Industries, and Forecast Industries) are working to get prototypes tested while starting to solicit commercial orders. Still another three (Alcoa, Rockwell International, and McDonnell Douglas) have com-

Figure 2-22. DAF-Indal diesel integrated VAWT, located at Sudbury, Ontario.

pleted research efforts but are not yet offering hardware on a commercial basis. (See Table 2-2.)

DAF-Indal, Ltd. The major thrust of the DAF commercialization effort has shifted to uprated versions of their 120 ft high by 80 ft diameter two-bladed VAWT which was originally prototyped in 1977 as a 230 kW machine on the Magdalen Islands in the Gulf of St. Lawrence. The new design has been rated at 500 kW by increasing rotational velocity and blade size. (See Figure 2-23.)

Table 2-2. Current and Expected Producers of Intermediate Size WECS (76 kW to 750 kW).

| WECS PRODUCER | POWER (kW) | | ROTOR | | | |
	CAPACITY	RATED (WINDSPEED)	SIZE	SPEED	BLADES	TYPE
DAF-Indal	600	500 (40 mph)	120 ft × 80 ft	45 rpm	2	Darrieus VAWT
Westinghouse	500	500 (22.5 mph)	125 ft diam.	42 rpm	2	downwind HAWT
WTG Energy Systems	350	200 (30 mph)	80 ft diam.	30 rpm	3	upwind HAWT
WTG Energy Systems	750	600 (30 mph)	125 ft diam.	25 rpm	3	upwind HAWT
Carter Wind Systems	150	125 (30.5 mph)	64 ft diam.	75 rpm	2	downwind HAWT
California Energy Group	130	125 (30 mph)	64 ft diam.	52 rpm	3	downwind HAWT
Flow Industries	120	100 (30 mph)	75 ft × 55 ft	50 rpm	2	Darrieus VAWT
Forecast Industries	188	185 (37 mph)	82 ft × 60 ft	48 rpm	2	Darrieus VAWT

Figure 2-23. DAF-Indal 230/500 kW VAWT. Photo is of 230 kW prototype on Magdalen Islands. Production model has same rotor size but larger blades.

The Mississauga, Ontario based company has worked closely with Canada's National Research Council (NRC) and leading provincial hydroelectric companies since 1974. Under the leadership of Project Manager Chuck Wood, DAF has long been Canada's leading WECS company.

The Magdalen Islands prototype was funded by NRC and Quebec Hydro and was installed and operated by Quebec Hydro's Institute of Research (IREQ). That original prototype spun out of control and was destroyed in mid 1978. The prototype was rebuilt, reinstalled and began generating electricity again in March, 1980. DAF's 230 kW VAWT utilized 24" chord extruded aluminum airfoils and operated at 36.5 rpm to drive a 300 kW three-phase induction generator. The new version maintains the same rotor size but utilizes Alcoa's 29" chord extruded aluminum blades for a 21% increase in solidity. Rotor operating velocity has been increased to 45 rpm to make the VAWT more productive in higher wind regimes.

Initial deliveries of the new 500 kW DAF VAWTs are scheduled for late 1982 for windfarm projects in California.

Westinghouse Electric Corporation. Based on its experience as contractor for DOE/NASA's 200 kW MOD-0A (prototypes were built and are currently being tested at Clayton, New Mexico; Block Island, Rhode Island; Culebra Island, Puerto Rico; and Kahuku Point on Oahu, Hawaii) Westinghouse concluded that HAWTs in the intermediate size range have advantages over both smaller and larger WECS. As a result, they responded to experience with the four MOD-0As and developed an improved version rated 500 kW in 22.5 to 44 mph wind velocities (measured at 30 ft). (See Figure 2-24.)

The new machine has been designated the Westinghouse WWG-0500. Like the MOD-0A, it has two variable pitch downwind blades which sweep an area with a 125 ft diameter. The new machine rotates at a higher speed, 42 rpm, and utilizes wood blades. It is available with either a synchronous or an induction generator to provide 4160 volt, three-phase, 60 Hz electricity. A 93 ft high pipe truss tower provides a hub height of 100 ft and blade tip ground clearance of 37 ft. Access is provided by a hoist located in the tower. The new HAWT utilizes a planetary gearbox instead of the conventional three-stage speed increaser specified in MOD-0A.

As operational experience with the MOD-0As provided data and confidence to Westinghouse, the new design evolved and sales orders were solicited from utilities and windfarmers. Project manager Will Treese indicated that orders for at least 100 units were needed to justify a production run and allow reasonable pricing. In March, 1982 that goal was achieved and the first of a series of projects was announced.

That first project is a windfarm for Southern California Edison (SCE) in the San Gorgonio Pass, approximately ten miles northeast of Palm Springs. First National Corporation (Birmingham, Michigan) and Manley, Bennett, McDonald and Company (an investment banking firm) formed a joint venture to serve as the small power producer to install and operate 60 WWG-0500s and sell the electrical output to SCE. The 30 MW windfarm is expected to be under construction

Figure 2-24. Westinghouse 200/500 kW HAWT. Photo is of 200 kW MOD-OA built by Westinghouse for DOE/NASA. Production model will utilize higher rotor speed to provide 500 kW rating.

by the end of 1982 with the first units on-line by the end of 1983. The construction project is planned for completion by early 1985 and sale of the electricity is expected to be contracted with SCE for a period of 15 to 30 years.

WTG Energy Systems, Inc. Since 1975, when it was incorporated specifically to design and manufacture commercial wind turbine generators, WTG has evolved to be the leading producer of WECS with capacities in excess of 100 kW. The Buffalo, New York based company, working independently of government programs, installed four of its Model MP-200 in North America and has a fifth unit under construction in South Wales, England. In addition to MP-200, WTG is now offering Model MP-600.

WTG is unique in the way it rates the power of its HAWTs. Although MP-200 has consistently generated 300–350 kW in the 35–50 mph wind velocity range (and in test conditions has generated as much as 400 kW) WTG's Director of Marketing, Al Gross, has insisted on the 200 kW rating and, along with Alcoa, has encouraged comparison of machines only on the basis of "cost per annual kWh" to take into consideration operating characteristics of individual WECS in specific wind regimes at projected installed cost of the hardware. MP-600, WTG's "600 kW" HAWT, is expected to generate peak power in excess of 750 kW.

The prototype of Model MP-200, MP1-200 was installed in July, 1977 on Cuttyhunk Island, Massachusetts and has completed a three year operational testing and economic evaluation program. That prototype utilized a synchronous generator so that it could be interconnected with the island's small 465 kW grid and its existing diesel generators. Several modifications have been made to the original prototype but the machine has performed well and has been providing useful electricity to Cuttyhunk's town of Gosnold. (See Figure 2-25.)

The first commercial unit, MP2-200, was sold to the Nova Scotia Power Corporation in the fall of 1979 and that machine generated its first power in February, 1981 at the Wreck Cove Hydro Plant. That unit is utilized to pump water for higher head hydro storage.

MP3-200 was sold to Pacific Power and Light Company (PP&L) and installed at that utility's WECS research site at Whiskey Run

Figure 2-25. WTG Cuttyhunk prototype.

on the windy Oregon coast. It generated its first electricity in January, 1981. Although originally installed with a synchronous generator, an induction generator was later substituted to make the WECS more compatible with the utility grid. The machine was turned over to PP&L for routine on-line operation. (See Figure 2-26.)

The fourth unit, MP4-200, was installed on Little Equinox Mountain in western Vermont as a mini-windfarm for delivery of electricity to Central Vermont Public Service Corporation in December, 1981. (See Figure 2-27.) The fifth unit, MP5-200, is scheduled for installation at a Carmarthen Bay site in South Wales, England for the Central Electricity Generating Board in late 1982.

Model MP-200 has three upwind 40 ft long blades (80 ft diameter swept area) which operate at 30 rpm to drive either a synchronous or an induction generator. WTG recommends the synchronous system for stand alone (less than 800 kW utility systems) installations and the induction system for "infinite bus applications."

WTG's MP-600 has a 125 ft diameter rotor with three upwind blades which operate at 25 rpm to drive a 750 kW induction generator intended to deliver 4160 volt, three-phase, 60 Hz utility compatible electricity. A 110 ft high tower is planned.

Potential Producers of Intermediate Size WECS. Three additional companies (California Energy Group, Flow Industries, and Forecast Industries) have announced their intention to develop and commercialize WECS in the 100 kW size range. Another three (Alcoa with its 100 kW and 500 kW capacity VAWTs, Rockwell International with its ill-fated 600 kW DOE/NASA MOD-6H HAWT, and McDonnell Douglas with a 250 kW version of its DOE/Rocky Flats straight-bladed 40 kW VAWT) have their machines on hold.

California Energy Group, Inc. is a new Santa Ana, California based company which has utilized the resources of Southern California aerospace and meteorological companies to design what it describes as a "utility class wind turbine generator," the Turbowind 64. It announced in March, 1982 that it planned a prototype installation later in the year. Its "design and engineering check list" described a HAWT with three downwind rotor blades with a diameter of 64 ft. The variable pitch blades were planned to operate at 50 to 55 rpm to drive a 150 hp induction generator to provide either 50 Hz or 60 Hz AC electricity at 480 volts.

Flow Industries, Inc. is a minority owned multinational company headquartered in Kent, Washington. It assembled a team in 1981 to

Figure 2-26. WTG 200 kW HAWT at Whiskey Run.

Figure 2-27. WTG 200 kW HAWT in Vermont.

form FloWind Corporation to develop and commercialize a "third generation" version of the DOE/Sandia 17 metre VAWT. The Flo-Wind Model 100 has a smaller rotor (75 ft x 55 ft vs 82 ft x 55 ft) than the "low cost 17 metre VAWT" built at three locations for DOE/Sandia by Alcoa. The Model 100 rotor is higher off the ground and the rotor turns at a higher rpm, enabling greater energy capture.

FloWind has stressed keeping components small to fit into standard shipping containers, protecting against salt spray corrosion, and survival in very high storm winds (up to 200 mph) with the objective of serving the world market, especially the wind-rich Pacific Islands. It has simplified installation through use of a hydraulic tilt-up system so that a heavy crane is not needed.

Although FloWind is new to the WECS industry, Flow Industries is an established high technology company with many of the engineering skills needed for WECS development. Flow is the world leader in the development of commercial waterjet cutting systems.

FloWind built a prototype of its 120 kW VAWT design at Black Angus Ranch near Ellensburg, Washington in early 1982 and has started to solicit sales orders. (See Figure 2-28.)

Forecast Industries, Inc. is a start-up company in Albuquerque, New Mexico which is also developing a "third generation" version of the DOE/Sandia 17 metre VAWT. However, Forecast is targeting its version for the domestic United States market and is emphasizing low cost, reliability and safety in its development program. (The author has to like this one. He is the founder and president of Forecast.)

The first Forecast model is designated VAWTPOWER 185 and is rated 185 kW at approximately 37 mph. Rotor height is slightly shorter than the DOE/Sandia/Alcoa "low cost 17 metre VAWT," at 82 ft, but its diameter is slightly larger (60 ft) to provide increased swept area. The basic rotor design has been simplified and strengthened by incorporating Alcoa's 29" chord extruded aluminum airfoils (increased from the 24" chord on the DOE prototypes). The 21% increase in solidity, coupled with the larger swept area, accounts for the rated power increase from 100 kW to 185 kW. (See Figure 2-29.)

Forecast is an Alcoa licensee of technology for small VAWTs (with Alcoa's 300/500 kW model specifically excluded) and, like Alcoa and Flow, has a technology transfer agreement with Sandia National Laboratories.

The Producers of Large WECS (Larger than 750 kW)

Although there was broad acceptance of DOE's assumption in the 1970s that bigger is better, only the Hamilton Standard Division, United Technologies Corporation, appears to be pursuing large WECS without direct government funding. (See Table 2-3.)

Bendix Corporation made an effort in the '70s with an innovative three-bladed HAWT which was originally designed and prototyped

Figure 2-28. Flow Model 100 VAWT prototype.

Figure 2-29. Forecast VAWTPOWER 185. Photo is of Alcoa 60 kW prototype on which the Forecast model is based.

(at a smaller scale) by Charles Schachle's Wind Power Products, Inc. (See Figure 2-30.) Bendix built a research prototype of a 3 MW capacity version at Southern California Edison's WECS research center in San Gorgonio Pass in 1980. (See Figure 2-31.) However, when that machine did not appear promising Bendix switched to

Table 2-3. Current Producer of Large WECS (Over 750 kW).

WECS PRODUCER	POWER (kW)		ROTOR			TYPE
	CAPACITY	RATED (WINDSPEED)	SIZE	SPEED	BLADES	
Hamilton Standard	4,000	4,000 (34 mph)	257 ft diam.	30 rpm	2	downwind HAWT

Figure 2-30. Original Schachle prototype, built by Wind Power Products in Moses Lake, Washington and later dismantled and shipped for a research project by New Mexico Engineering Research Institute.

Figure 2-31. Bendix/Schachle 3 MW HAWT, installed at SCE's Wind Energy Center and later retrofitted to a 1.3 MW capacity model.

another approach to large HAWTs and went back to the drawing boards to develop a more conventional HAWT in the 3–5 MW size range. Bendix has retrofitted their original prototype, and is now testing it as a 1.3 MW research machine.

Both the U.S. and Canadian governments are funding major large WECS R&D expected to result in technology which could be ready for commercialization by the mid- to late 1980s. DOE/NASA's 2.5

MW MOD-2 HAWT is furthest along with four Boeing-built research prototypes operating or being retrofitted and a fifth under construction for the Department of Interior's Bureau of Reclamation near Medicine Bow, Wyoming. DOE/NASA's MOD-5 HAWT was expected to take advantage of all the smaller MOD series designs and prototypes and evolve superior "third generation" HAWTs in the 6.2 MW (MOD-5A being designed by General Electric Co.) to 7.2 MW (MOD-5B being designed by Boeing Engineering and Construction) size range. The Canadian government, through NRC, is also concentrating its R&D funding on multi-megawatt size WECS and is proceeding with Project Aeolus to design and prototype a Darrieus type VAWT in the 3 MW to 4 MW size range.

Hamilton Standard Division, United Technologies Corporation. As the only North American producer of megawatt scale WECS, Hamilton Standard stands at the threshold of WECS industry leadership with two large prototypes nearing completion and numerous wind-farms on the drawing boards.

Hamilton Standard initiated its wind R&D program in 1973 with aerodynamic analysis and testing. That effort was expanded into system engineering studies and preliminary rotor system and blade designs during 1974 and 1975. The design tools which were initiated at that time have been continually refined and updated and are the backbone of Hamilton Standard's program today.

By the late 1970s, Hamilton Standard was concentrating on integration of rotors and blades with the overall wind turbines and focusing its R&D efforts on the development of improved high technology blades. When it couldn't get help from the U.S. government's large HAWT program, it initiated a program in Sweden to build and install a 3 MW HAWT in the town of Maglarp, near Malmo, for testing by Sweden's largest private utility, Sydkraft, under the auspices of the Swedish government. The development effort became essentially a joint veture with Swedyards (formerly Karlskronavarvet AB). Hamilton Standard concentrated on the design and manufacture (at a dedicated new 35,000 square ft manufacturing facility in East Granby, Connecticut) of filament wound glass fiber reinforced epoxy blades while their Swedish partners proceeded with the other portions of the project. The 3MW WTS-3 is expected to be ready for testing by mid 1982. (See Figure 2-32.)

Figure 2-32. Hamilton Standard's WTS-3 fabrication in the Swedyards plant in Sweden. Note scale compared to boat in background.

Based on its Swedish efforts, Hamilton Standard competed for and won a contract to supply a "system verification unit (SVU)" for the U.S. Department of Interior's Bureau of Reclamation (BuRec) for installation near Medicine Bow, Wyoming for integration into BuRec's giant hydroelectric grid. The 4 MW capacity BuRec model is more powerful than the Swedish unit and is designated WTS-4. That unit is also expected to be ready for testing by mid 1982. (See Figure 2-33.)

Model WTS-3 is a two bladed downwind HAWT with a rotor diameter of 257 ft, mounted on a 262 ft tall tubular steel tower, which drives a synchronous generator to deliver its 3 MW of 50 Hz power at approximately 31 mph (measured at the hub). The unit is expected

Figure 2-33. Hamilton Standard WTS-4: artists rendering of 4 MW HAWT being built for BuRec at Medicine Bow, Wyoming.

to operate in 10 to 50 mph winds and is designed to withstand storm winds of 130 mph. (See Figures 2–34 and 2–35.)

Model WTS-4 is also a two-bladed downwind HAWT with a rotor diameter of 257 ft. It operates at 30 rpm to drive a synchronous generator to provide its 4 MW of 60 Hz power at approximately 34 mph (measured at the hub). The 262 ft tower height is the same as for WTS-3 but details are different. The unit is expected to generate power in wind-velocities from 15 to 60 mph.

After having 20 of its WTS-4 machines chosen by Windfarms Ltd. for their 80 MW windfarm for Hawaiian Electric, and additional WTS-4 HAWTs specified for Windfarms Ltd's major project for Pacific Gas and Electric in Northern California, Hamilton Standard appeared in 1981 to be assuming a dominant position in the large WECS and windfarm markets. When those projects fell through, Hamilton Standard negotiated a more modest five unit 20 MW ca-

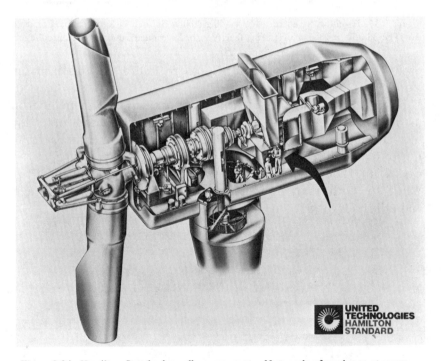

Figure 2-34. Hamilton Standard nacelle components. Note scale of workmen at arrow.

Figure 2-35. Hamilton Standard blade winding facility, located in East Granby, Connecticut. The 35,000 sq. ft. manufacturing facility is the largest dedicated to WECS production.

pacity windfarm of its own for Southern California Edison in San Gorgonio Pass.

Hamilton Standard has expended approximately $25 million on its ambitious wind energy program and believes it has evolved the most technically advanced HAWT design and manufacturing facilities to date. With its major investment and commitment to commercialization, Hamilton Standard plans to be providing cost-competitive WECS for both its own windfarms and for independent windfarmers and utilities by 1983.

3
Bulk Power:
Farming the Wind

If WECS are to make a substantial contribution to energy independence, reduction of imports, reducing balance-of-payments deficits, extending the life of depletable natural resources, or, most importantly, to saving or making money, they must be installed in quantity and provide bulk power.

Large sites with good to great winds have to be utilized. Organizations with the financial, technical, and marketing capability to fund, engineer, build, and operate major projects and market and distribute the generated electricity must be involved.

FRIENDLY LOCAL UTILITIES

Although others have been suggested and many have tried, it is now clear that electrical power is the business of the nation's friendly local utilities. And no matter what role others may play, somewhere along the line the utilities must be involved in the generation, transmission or distribution functions, or in all of them.

The local utilities fall into three general classes—investor owned utilities, publicly owned utilities, and rural electric cooperatives. All have their own problems and advantages and disadvantages relative to WECS. They all have common problems with regulators, in raising capital, and in dealing with politicians. The electric utility industry has not been fun since the oil and nuclear problems of the 1970s. And with the capital intensity of all utilities, the high interest rates of the '80s have been less than helpful.

The electric power industry is the most capital intensive in the nation. According to Robert Mauro of the Electric Power Research

Institute, the average electric utility needs $5 in plant, property, and equipment to produce $1 in revenue. Steel needs $3 for every $1 while the auto industry requires $1 for every $1. Mauro stated, "The average cost for a new central generating plant of around one million kilowatts is $1600 for each kW. Coal plants cost less, nuclear plants cost more. In a little while, a one million kilowatt facility will cost $2000 a kilowatt."

Although most WECS projects require capital investments less than $2000 per kW and use free fuel, WECS do not produce their rated kW much of the time and can not necessarily be counted on for power when needed.

Integrating windpower into a utility system is not easy. The nearly constant variation in speed of the wind means that the wind turbine must be carefully controlled so that its electrical output is comparable (in phase, voltage, and frequency) with that of the utility system. Modern wind machines incorporate design features and control systems to ensure achievement of the match.

In addition to the features that might be designed into each wind turbine to ensure that it is electrically compatible, other issues may arise when numerous wind machines are connected to a utility system. The varying output of a few wind machines would be relatively easy to accommodate in a utility system since they would have an effect on system operation similar to that of large varying electrical demands. However, as the number of wind turbines becomes larger, a point could occur at which the utility system could not respond quickly enough to the variation. Ability to absorb the power fluctuations of the wind machines is determined by a number of factors including the utility size and the mix and size of individual generating units. In general, a utility with a large fraction of hydroelectric, pumped storage, or gas turbines should be able to accommodate a larger fraction of wind capacity. The modeling of utility system operation with wind capacity is an evolving science, and more operational experience is needed to improve understanding. Most large utilities should be able to profitably integrate WECS with a total of up to 30% of their generating power capacity, depending upon individual utility characteristics. Utilities can either build and own WECS or can contract with independent small power producers (SPP) for the purchase of electricity from their WECS generation facilities.

Although no utility relishes the loss of generation business, those that are short of baseload capacity or forced to make use of expensive oil and gas fired units are likely to see substantial benefits in supplementing their supplies. The availability of SPP WECS can defer the need for siting and building new utility plants. For capacity-short utilities, purchasing energy can be an economic course that helps relieve a cash squeeze by postponing costly new construction.

On the other hand, utilities that have adequate or ample capacity and can produce relatively inexpensive baseload electricity in coal-fired or nuclear plants tend to be less enthusiastic. They have no need for the SPP electricity, and they don't want to buy it at rates that are substantially higher than their own average generating costs.

Investor Owned Utilities

Investor owned utilities (IOUs) have probably had the most difficult time in adjusting to the new realities. Those businesses are the most directly affected by federal and state regulations and by political tampering. They also have the delicate job of satisfying the conflicting needs and desires of their customers, stockholders, and lenders for both the short and long term. Their capital requirements are immense. In 1980, investor owned utilities shelled out $27.3 billion for new facilities, absorbing around 14.5% of vital U.S. capital investment. From 1979 to 1984, the investor owned utilities expect to spend $155 billion, mostly in new generating capacity. It shouldn't be surprising that many IOUs have lost their entrepreneurial spirit and settled into a defensive posture while awaiting clarification of their marching orders.

Within that climate, the fact that unproven WECS have found their way into some IOU systems is almost amazing. Several IOUs are even spending some of their own scarce money to help develop or demonstrate WECS or to help promote use of WECS by their customers and construction of windfarms by small power producers. Among the industry leaders are Pacific Gas and Electric Company,

Southern California Edison Company, Pacific Power and Light Company, Pennsylvania Power and Light Company, Southwestern Public Service Company, Wisconsin Power and Light Company and Hawaiian Electric Company. Altogether, over 100 of the 218 investor owned utilities have some type of wind related programs.

West coast and Hawaiian utilities have been the most aggressive in working with WECS, probably because of practical and political limitations on other options and the fact that those areas are growing (and therefore need additional electricity capacity) and highly dependent on foreign oil and fully exploited water as fuels. Two major California utilities, Southern California Edison (SCE) and Pacific Gas and Electric (PG&E), are clearly the investor owned utilities most involved with WECS.

Southern California Edison has an active research program to develop and test WECS at its Wind Energy Center (see Figure 3-1) eight miles northwest of Palm Springs in the windswept San Gorgonio Pass.

Figure 3-1. Southern California Edison Wind Energy Center, located near Palm Springs in the San Gorgonio Pass.

The Wind Energy Center was dedicated in late 1980 and has been the site for hands-on R&D with the Bendix/Schachle 3 MW HAWT (currently being retrofitted by Bendix and SCE into a 1.3 MW prototype utilizing a MOD-1 gearbox and an induction generator), an Alcoa 500 kW VAWT (which self-destructed), and a DAF-Indal 50 kW VAWT, as well as detailed analysis of wind characteristics in one of California's best wind resource areas. (See Figures 3-2 through 3-5.) The SCE wind program was initiated in 1975 and has benefited from top level support by Chairman William R. Gould, Advanced Engineering Vice-President Larry Papay, and R&D Director Frank McCrackin. Bob Sheffler is SCE's wind projects manager with able assistance by program managers Mike Wehrey, Tony Fung, John Stolpe, and Bob Yinger. Bill

Figure 3-2. SCE 3 MW HAWT prototype by Bendix. Used as a research machine, this prototype has been modified several times and currently has 1.3 MW generating capacity.

Figure 3-3. SCE 0.5 MW VAWT prototype by Alcoa. This prototype was destroyed when it spun out of control to over 150% of its planned rotor speed on April 3, 1981.

Emrich has the responsibility of assisting small power producers and a job unique in the utility industry: project manager, wind commercialization.

In addition to direct R&D efforts, SCE initiated a unique "wind park" program in late 1980 by soliciting proposals from small power producers to supply electricity to SCE's grid (see Figure 3-6). There

Figure 3-4. SCE 50 kW VAWT prototype by DAF-Indal.

were 15 responses to that request for proposals and additional developers initiated unsolicited proposals later. By May, 1982 seven
wind parks were under development or had reached the point where
contracts had been executed between the small power producers and
SCE. When completed, those seven windfarms are expected to add
76.8 MW to 91.8 MW (depending on options) of power to SCE's total
electrical capacity. Contacts were in the process of negotiation with
17 additional small power producers whose wind parks, when completed, will add another 300 MW to 400 MW of power. Several of
the windfarms are discussed in greater detail later in this chapter.

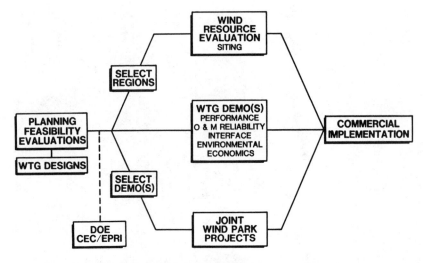

Figure 3-5. SCE total wind program.

Pacific Gas and Electric invested its own R&D funds in wind prospecting and site analysis and in the purchase and installation of one of the five DOE/NASA 2.5 MW MOD-2s fabricated by Boeing Engineering and Construction Company. (See Figures 3-7 through 3-11.) Their prototype was dedicated on July 15, 1982 on the Ruth King Ranch, in the windswept hills near Fairfield in Solano County, about 40 miles

● **OBJECTIVE: ACCELERATE INSTALLATION OF COMMERCIAL WIND PARKS**
- ● PRIVATE DEVELOPERS OWN THE WIND PARK AND SELL ELECTRICITY TO EDISON
- ● EDISON IS WILLING TO NEGOTIATE ON DEGREE OF PROJECT PARTICIPATION AND RISK ASSUMPTION
- ● EDISON PAYS LESS THAN PURPA AVOIDED COST

● **BENEFIT TO EDISON**
- ● MINIMIZE TECHNICAL RISK AND CAPITAL REQUIREMENTS
- ● DECREASE DEPENDENCE ON OIL
- ● HELP MEET ELECTRICAL GENERATION NEEDS

● **BENEFIT TO PRIVATE DEVELOPERS**
- ● UTILIZE TAX CREDITS AND ACCELERATED DEPRECIATION TO ACHIEVE PROFIT LEVELS COMMENSURATE WITH THEIR RISKS

Figure 3-6. SCE wind park project.

Figure 3-7. Pacific Gas and Electric MOD-2 prototype, built by Boeing Engineering and Construction Company.

northeast of San Francisco. PG&E hopes to get over seven million kWh per year out of the big HAWT while further analyzing performance of the machine, the area's wind characteristics, and the interaction between its grid and the wind turbine.

Development of northern California's wind energy resources is a natural evolution for PG&E as it gets experience with WECS hardware and a better understanding of the extent and characteristics of the windpower available. PG&E has accepted the conclusion that wind can be an important part of its total energy system by the turn of the century. From Chairman Frederick W. Mielke, Jr. and President Barton W. Shackelford throughout the company, they are giving WECS a chance to demonstrate credibility.

In addition to the major expenditure for the purchase and operation of the MOD-2, PG&E participated with the California Depart-

Figure 3-8. PG&E MOD-2 blade ready for lift.

ment of Water Resources in a pioneering VAWT research project at the San Luis Reservoir in Merced County. That 50 kW VAWT, supplied by DAF-Indal, has been operating since March, 1981.

Another several dozen small WECS have been interconnected to PG&E lines and some of the electricity generated by those machines is being purchased by the utility. On a larger scale, two windfarms are in operation in the Altamont Pass, east of Livermore in Alameda

Figure 3-9. PG&E MOD-2 blade installation.

County. U.S. Windpower, Inc. had 100 of its 50 kW capacity HAWTs feeding electricity to PG&E by the spring of 1982, and was gearing up for another 100 on an adjacent site and an additional 400 down the road, for a total of 30 MW. Fayette Manufacturing Co. had 34 of its 50 kW HAWTs up by spring on a site adjacent to the first two windfarms by U.S. Windpower, and was proceeding toward a total of 300, or 15 MW of power capacity from which energy will be sold to PG&E. The wind resources in the Altamont Pass have attracted additional developers who are working on windfarms of their own. PG&E is in the process of extending their transmission lines and increasing their capacity to accept power from those small power producers.

Figure 3-10. PG&E MOD-2 near completion.

Although not yet under construction, the world's largest windfarm is being planned as a cooperative venture between PG&E, the California Department of Water Resources, and Windfarms, Ltd, the San Francisco based developer. (See Figure 3-12.) Announced in April, 1981, that project anticipates a total generating capacity of 350 MW. Installation of the wind machines is planned in three stages over a period of seven years. The Solano County site's wind resources are being counted on to generate almost a billion kWh of energy per year when the project is on line.

Figure 3-11. PG&E MOD-2 prototype ready for action.

Credit should be given to Peter Hindley for getting the total PG&E wind program started in the 1970s. Paul Petersen has been the engineer in charge of construction and operation of the MOD-2. Nolan Daines, Vice-President of Planning and Research, and engineers John Wells, Tom Hillesland, and Hal Seielstad have been active in keeping PG&E on a positive course toward utilizing northern California's wind resources.

Among the other investor owned utilities, Portland based Pacific Power and Light Company has had an active wind prospecting program and participated in formulating the state of Oregon's impressive wind energy program. In 1981, PP&L developed a WECS research site at Whiskey Run, on the Oregon Coast, and installed one of WTG Energy System's 200 kW HAWTs (see Figure 3-13) as the first of several planned research machines.

Figure 3-12. Planned world's largest windfarm. PG&E, California Department of Water Resources, and Windfarms, Ltd., are cooperating to develop this 350 MW project in Solano County, near San Francisco. (Artist's sketch.)

Pennsylvania Power and Light Company was one of the first utilities to get involved with WECS. During the mid 1970s it worked with the late Terry Mehrkam in helping to develop Mehrkam Energy Development Corporation's 40 kW, 100 kW, and 225 kW prototype HAWTs in central Pennsylvania. Metropolitan Edison Company also worked with Mehrkam in prototyping his ill-fated 2 MW HAWT for Metals Engineering Company in Leesport, Pennsylvania in 1980.

Wisconsin Power and Light, under Program Manager Carel De Winkel, has had an aggressive program to encourage installation of small WECS on dairy farms and in residential applications for test. Their program had unusually bad luck with early installations in 1980. Three 10 kW capacity HAWTs (two DOE research prototypes by United Technologies Corporation and an early Jacobs unit) were destroyed in storms late that year. Still another unit, an early Carter 25 kW model, had problems with a gearbox and had to be reinstalled before it operated successfully. However, those pioneering installations led to many additional WECS projects in Wisconsin.

Hawaiian Electric Company has given special attention to WECS because of its powerful Pacific winds and almost total dependence on oil for its electricity generation. It has encouraged and supported many small WECS installations on Oahu, is the host utility for the most productive of the four DOE/NASA 200 kW MOD-0A HAWTs, and has been actively participating in project planning and site anal-

Figure 3-13. Pacific Power and Light's 200 kW HAWT prototype, located on the Oregon Coast at Whiskey Run, supplied by WTG Energy Systems.

ysis with Windfarms, Ltd. and its consultants in preparation for the announced 80 MW windfarm at Kahuku Point on Oahu's north shore.

The Electric Power Research Institute (EPRI), which attempts to monitor wind energy activities by utilities, estimates that over 100 utilities have some form of wind energy program. As many as 40 may be considering purchases of WECS.

Publicly Owned Utilities

Some of the publicly owned utilities are among the largest generators of electricity in the world. Others are very small in size and only distribute electricity which they purchase from others. In 1981 there were 2206 local publicly owned systems (including municipals, public utility districts, state systems and county systems), 51 publicly owned joint action agencies, and eight federal power agencies. Of those 2265 utilities, approximately two-thirds purchase all or part of their power at wholesale from others. Over 900 of those buy all or part of their power from investor owned utilities.

Where the wind resources are adequate, all of these utilities can benefit from WECS, although some will be better served purchasing power from small power producers or other WECS owners on their lines. The utilities that distribute only are looking at WECS as energy savers which can reduce their purchased cost of electricity. The utilities which generate are looking at WECS as supplements to their hydroelectric and diesel generating systems where they can be valued based on both fuel savings and capacity credit.

Of the 675 local public power systems in the United States which generate at least some of their own electricity, 500 have diesel engine sets and 172 have steam plants. With the last decade's rapid escalation of prices of both natural gas and oil, and the planned deregulation of natural gas prices in the mid-1980s, the cost of generating with that equipment is making investments in WECS increasingly attractive.

The giant federal bulk power producers and marketers have the potential for getting the most value out of WECS and making the greatest contribution to the nation's energy needs. Several have research, development, and demonstration projects underway. The Department of Interior's Bureau of Reclamation (BuRec) has two large WECS (DOE/NASA 2.5 MW MOD-2 and Hamilton Standard's

4 MW WTS-4) at their windfarm site near Medicine Bow, Wyoming (see Figure 3-14) and is testing those machines as "system verification units" to determine if BuRec should proceed with one or more large windfarms in the windy western states which they serve. Tennessee Valley Authority (TVA), although it has access to much less wind in its portion of the country, has prospected for available winds, published wind maps for their territory, and has encouraged windfarms in the few windswept mountain areas that have been identified. Bonneville Power Administration (BPA) has been active with WECS in the windy Northwest both as operator of the DOE/NASA 7.5 MW experimental windfarm in the Klickitat County Public Utility District on the Washington side of the Columbia River Gorge and as encourager of many other WECS tied to its power lines or the lines of its local utility customers.

On a smaller, but no less important scale, individual public utilities have proceeded on their own, or in cooperation with the American Public Power Association (APPA) or the Department of Energy (DOE) to get WECS projects up and operating so that meaningful evaluations and future plans can be made.

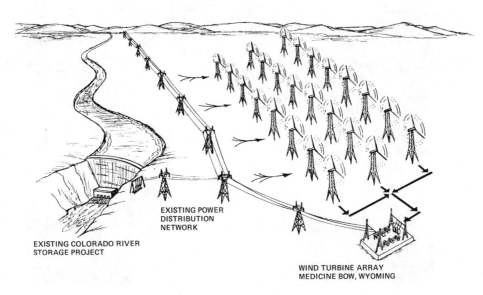

EXISTING POWER
DISTRIBUTION
NETWORK

EXISTING COLORADO RIVER
STORAGE PROJECT

WIND TURBINE ARRAY
MEDICINE BOW, WYOMING

Figure 3-14. BuRec's Medicine Bow WECS project. Hamilton Standard's WTS-4 and DOE/ NASA's MOD-2 are being built and tested as system verification units.

The California Department of Water Resources, Sacramento, installed a 50 kW VAWT (by DAF-Indal) at their San Luis Reservoir about 70 miles east of San Francisco. That proof-of-concept experiment, along with wind resources analysis, is intended to serve as the starting point for addition of over 100 MW of windpowered generating capacity by the year 2000. (See Figure 3-15.)

Eugene Water and Electric Board (EWEB, in Eugene, Oregon) and Central Lincoln Peoples Utility District (CLPUD, in Newport, Oregon) organized a total of 17 municipals, People's Utility Districts (PUDs) and Rural Electric Cooperatives (RECs) and secured a small APPA research grant to proceed with installation of the world's largest VAWT (500 kW capacity by Alcoa) at a windy site above Agate Beach on the Oregon coast. (See Figure 3-16.) Installed early in 1981, that five-year joint research project was intended to test the value, reliability, costs, and impact on the utility system of WECS.

The Power Authority of the State of New York (PASNY) surveyed 40 sites before selecting seven for detailed analysis while proceeding toward their first WECS installations.

Figure 3-15. California Department of Water Resources 50 kW VAWT prototype; built in cooperation with PG&E, supplied by DAF-Indal.

Figure 3-16. Oregon coast 500 kW VAWT prototype; developed by Eugene Water and Electric Board and Central Lincoln Peoples Utility District at Agate Beach, supplied by Alcoa.

The city of Santa Clara is working on a WECS project in one of their nearby mountain ranges. Princeton, Massachusetts surveyed their wind resources in anticipation of installing a 500 kW WECS. Lincoln, Nebraska bought a small WECS and interconnected it with their municipal electric distribution system. Washington Public Power Supply System (WPPSS) is studying the potential of adding WECS to its generation base in eastern Washington. Atlantic City Electric Company was the first municipal utility to offer cash incentives ($500) to customers who install WECS by September, 1983. Three of the DOE/NASA 200 kW MOD-0A research machines are hosted by publicly owned utilities (City of Clayton, New Mexico; Block Island, Rhode Island; and the Puerto Rico Electric Power Authority).

Rural Electric Cooperatives

The nation's 916 Rural Electric Cooperatives (RECs) serve approximately 25 million people and provide electricity to about 75% of the land area of the United States. Only Connecticut, Hawaii, Massachusetts, and Rhode Island have not taken advantage of the low interest loans and other benefits since the Rural Electrification Administration (REA) was created in 1935 and given access to the U.S. Treasury's Federal Financing Bank in 1937.

All of the RECs are member-owned, non-profit organizations, but only 46 of the total are in the generation business. The remaining 870 are distribution cooperatives which buy their electricity from others and pass it on to their members.

RECs serve many of the windiest portions of the U.S. and, because of their rural nature, both the cooperatives and their individual members have open land where effective WECS can be sited. Many farms and rural residences now served by RECs previously utilized water pumping windmills and battery charging windchargers prior to creation of the REA. As a result, individual cooperatives and the National Rural Electric Cooperative Association (NRECA) have taken an active role in the revitalization of WECS since the mid 1970s.

NRECA, under Lowell Endahl and Wilson Prichett, has concentrated on the issues of safety and quality of service, particularly as they relate to the interconnections and interrelationships between WECS producers, WECS owners (often cooperative members), the local cooperative (which may also be a WECS owner), and the generating and transmitting cooperative or supplier utility. NRECA co-sponsored a few WECS R&D projects to develop data and has also actively monitored many of the local WECS installations by RECs or tied into cooperative lines. From that experience, NRECA has made a series of recommendations to its members.

In addition to advising members on WECS issues such as economics, insurance, and compliance with relative laws and regulations, NRECA has attempted to provide its members with an understanding of WECS operating characteristics and ways that the cooperatives can better deal with specific issues and be of maximum assistance when WECS are installed.

NRECA has primarily concentrated on issues of interconnection between WECS and the cooperative lines and how the cooperatives can best accommodate dispersed WECS in their systems. The main safety and quality of service areas with which NRECA has been of assistance are:

- Electric line and equipment protection;
- Quality of power;
- Energy metering.

NRECA's extensive investigations of all of these issues found none to be unique to WECS and none which were expected to be major problems for the nation's Rural Electric Cooperatives if modern proven WECS hardware utilizing induction or synchronous AC generators, or DC generators with properly engineered and built inverters, are utilized. In the unusual circumstances where special treatment of a WECS installation is needed, solutions are known and hardware to allow the local system to function properly with the addition of WECS is available. Many helpful publications are available from NRECA if specific problems need attention.

Local RECs have been active with WECS projects and hundreds of WECS are known to be interconnected with cooperative lines in many parts of the country.

The Blue Ridge Electric Membership Cooperative, headquartered in Lenoir, North Carolina, hosted DOE/NASA's 2 MW capacity MOD-1 on Howard's Knob in the mountains overlooking Boone, North Carolina. (See Figure 3-17.) Construction of that research machine, which was the world's largest at the time, started in June, 1979. Although that government prototype was technically unsuccessful, much was learned which has been helpful to later generations of large WECS. Blue Ridge is cooperating with DOE/NASA and General Electric (the prototype's builder) to retrofit or otherwise make the most out of the research project.

Allegheny Electric Cooperative, the generating and transmitting coop for Pennsylvania's distributing cooperatives, developed two WECS research projects in cooperation with NRECA, the Pennsylvania Rural Electric Cooperatives Association, and two of its distributing coops, Southwest Central Rural Electric Cooperative Corp. in Indiana, PA and Adams Electric Cooperative Corp. in Gettysburg. As with the case of the government's MOD-1 project in North Carolina, the WECS hardware was not technically successful and little power was generated. However, the experience was valuable in determining wind characteristics and establishing procedures for installing, interconnecting, and operating WECS on a cooperative

Figure 3-17. Blue Ridge 2 MW MOD-1 prototype, installed on a mountain above Boone, NC; built for DOE/NASA by General Electric.

basis by the generating cooperative, distributing cooperatives, and members or customers with sites.

The East River Electric Power Cooperative, Inc., headquartered in Madison, South Dakota, has been monitoring and analyzing winds in its service area since 1976. East River, which is the generation and transmission cooperative serving 24 distribution coops in South Dakota and one in Minnesota, has located a site on a hill near Huron, South Dakota (also near the Beadle Electric Cooperative's headquarters building) where winds average close to 17 mph. East River is hoping to get one of DOE/NASA's giant MOD-5s for installation and test at that site.

Swisher Electric Cooperative, headquartered in Tulia in the windy panhandle area of Texas, has had years of exposure to traditional water pumping windmills in its service area and has long been aware of the potential of wind energy. Since 1979, it has cooperated with the State of Texas, West Texas State University, and Wind Engineering Corporation in developing an innovative 25 kW WECS project intended to provide both electricity and a mechanical assist for irrigation pumping. (See Figure 3-18.) Within Swisher's territory, irrigation pumping has been by diesel, gasoline, and natural gas powered internal combustion engines, as well as electric motors, usually in the 20–100 kW power range. With the escalating costs of those options, and with threatened shut-offs or deregulation of natural gas for irrigation purposes, the concept of utilizing wind energy for irrigation pumping, as well as livestock watering stations, has attracted considerable interest.

The nation's friendly local utilities have responded to the WECS opportunity in a variety of ways ranging from total disregard to enthusiastic support and purchase. Some utilities have concentrated on hardware R&D, some have supported R&D efforts by third parties, a few have purchased WECS hardware to get hands-on experience, a few have encouraged windfarms on their lines to augment their own generation, and many more have taken a position of waiting and watching while the industry leaders work their way through the problems and expenses of pioneering new technology.

There are many reasons why all utilities which have windswept land within their service areas should have an active interest in wind. In addition to the obvious advantages of utilizing locally

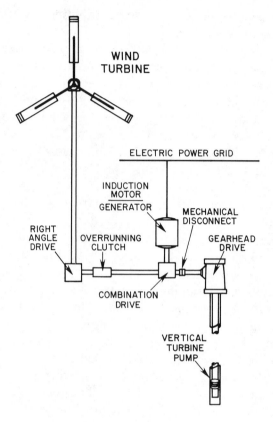

Figure 3-18. (a) Swisher Electric's HAWT irrigation project. Schematic design of WECS mechanically augmenting electric power for irrigation pumping.

available, nondepletable, free fuel, the wind is relatively clean, non-polluting, and without the need for either water for cooling or a means to dispose of waste. Less obvious are the advantages of high energy gain (energy out of a WECS compared to energy needed to build and fuel the WECS is better than any other known generator) and high energy yield per land utilized (properly sited WECS clusters can generate more energy per acre of dedicated land than any other power system except nuclear power plants).

However, the major advantages of WECS are in economics and the fact that scarce utility dollars can be preserved by having third party participants (small power producers) build the WECS power plants

Figure 3-18. (b) The Wingen 25 installed near Tulia, Texas to demonstrate the Swisher irrigation concept.

and supply electricity to the utilities for marketing, transmission and distribution to the ultimate users. When interest rates become affordable, or when utilities can again save up enough money or reestablish their credit, they can build their own windfarms in incremental or modular sizes to respond quickly to customer demand and for replacement of aging facilities.

Because of the intermittent availability of the wind, WECS can't be the total generating capacity of a utility unless it can purchase electricity on demand from others, but there is no reason why WECS can't economically provide 10–30% of a utility's total generating needs if the utility has adequate sites with powerful winds.

PURPA, FERC, AND THE REGULATORY THRUST

With recognition that WECS could not make a contribution to the nation's energy needs without active involvement by the electric utilities, congress passed the Public Utility Regulatory Policies Act of 1978 (PURPA), which directed that the Federal Energy Regulatory Commission (FERC) work with State Public Utility Commissions (PUCs) and the utility industry to make sure that WECS and other dispersed generators of electricity be given a chance.

PURPA was intended to create a market for power generated by Small Power Producers (SPP), including owner-operators of WECS, and cogenerators. It required utilities to gather and make public information about system costs and planning. It tried to reduce the amount of regulation, paperwork and delays in the relationships between utilities and the third-party generators. Most importantly, PURPA intended that utilities have to allow SPPs to interconnect with their lines and have to pay a favorable rate for electricity fed into the grid and to sell electricity, if needed, at a reasonable rate to the SPP or cogenerator.

The idea of PURPA was sound and WECS participants welcomed and supported it. However, imprecise wording in the law and the inevitable delays in interpreting and implementing the law provided the opportunity for a few states to fight back on the grounds of states rights and the fact that they didn't like being told by the federal government how to regulate utilities in their states. Most utilities took the opportunity caused by the bureaucratic infighting and confusion to delay establishing their rules, regulations, and rate structures for both sale and buy-back of electricity. In some of the states which did move forward with implementation of the law, if the utilities didn't like the interpretations, they took advantage of the federal–state bickering and lack of leadership by FERC to challenge the state's rules and buy-back rates.

By early 1982 the future of important features of PURPA was in doubt. Court suits and rulings led to additional challenges and appeals.

Key congressmen were attempting to understand what needed to be changed or strengthened in the PURPA law so they could introduce legislation to get it back on track. However, potential new legislation is further complicated by the desire of the utilities themselves to be able to own small power production or cogeneration facilities and take advantage of many of PURPA's features.

Regulatory problems in the utility industry are not unique to wind energy or WECS. What is described as "the regulatory problem" is considered by many to be the most important obstacle to orderly growth to meet the nation's energy needs. Caught up in the political pressures of the times, few regulatory agencies have the competence or courage to take the actions needed to allow the utilities to move forward with integrated plans which will allow utilities to make adequate money to invest in the proper mix of generating, transmitting, and distributing facilities while providing "least cost" electricity to their customers.

In addition to the basic controversy relative to who has the right to regulate utilities, there are two main features of PURPA being argued. One is the definition of a fair and reasonable price for the utilities to pay for the electricity to be fed back into their lines. The other is the mandate that FERC grant blanket authority for SPPs and cogenerators to interconnect with utilities without meeting other federal requirements. When all is said and done, the main open issue relative to WECS is the price that WECS can command for their electricity sold to the utility.

Some of the PURPA features not challenged include establishing standard rates for facilities with less than 100 kW capacity, exempting from federal and state regulation facilities (such as windfarms) with 30 MW or less of generating capacity, and to regulate facilities in the 30 MW to 80 MW size range under special provisions of PURPA, instead of more cumbersome regulation imposed on utilities.

Once congress, FERC, and the states get the PURPA act together and favorable rate structures are established in the windy areas, WECS projects will be able to move forward. Individual dispersed units, as well as clusters of units totaling up to 80 MW of capacity, can be built to fulfill the objectives of congress and start contributing to the energy supply mix.

SMALL POWER PRODUCERS IN ACTION

The many government programs and incentives created to accelerate WECS commercialization have been successful in attracting third-party organizations to develop WECS projects and sell the WECS power to utilities. Such organizations have been called Independent Power Producers (IPP) by some, Interim Risk Absorbing Companies (IRAC) by DOE, and windfarmers by most. However, for purposes of gaining the most favorable treatment under PURPA, the term Small Power Producer (SPP) is most appropriate.

In addition to PURPA, the crude oil Windfall Profits Tax Act of 1980 (which provided 15% tax credits, in addition to the regular 10% investment tax credit, for investments in alternative energy equipment such as WECS), the ambitious but never funded Wind Energy Systems Act of 1980, and the important Economic Recovery Tax Act (ERTA) of 1981 (which provided accelerated five year depreciation of equipment such as WECS) combined with initial signs of encouraging results with WECS hardware to give birth to a dynamic windfarming industry. The incentives proved powerful and irresistable. They created a new and immediate market for WECS with the new small power producers as well as augmenting more traditional WECS markets.

When combined with liberal state tax credits and other incentives, the temptation to develop a windfarm and become a SPP is almost too strong to resist. And potential economic rewards are almost too attractive to be believed. California, for instance, has utilized the "carrot and stick" approach by providing liberal state incentives to SPPs (the carrot) while using its state Public Utility Commission to force utilities, under threat of penalties (the stick), to take an aggressive position in encouraging and cooperating with small power producers. California incentives to SPPs include an additional 25% state tax credit, accelerated state depreciation instead of the tax credits or for the amount above the tax credits, low interest loans by a special Alternative Energy Financing Authority, and loans with Small Business Administration (SBA) guarantees by the California Business and Industrial Development Corporation. Not surprisingly, California has attracted most of the nation's WECS projects and is on its way to meeting the state goal of having 10% of California's energy generated by the wind by the year 2000.

Of course there is danger in such rapid development of windfarms which, almost without exception, have to utilize unproven WECS hardware and inexperienced people. Early failures of hardware or projects will tend to dampen enthusiasm and cause governments and regulatory agencies to react and perhaps overrespond by withdrawing incentives or forcing unreasonable standards. Much is at stake as the pioneering windfarms evolve and get some running time in California and elsewhere.

The small power producer phenomenon has evolved to fill a void between the utilities and the WECS producers. Both of those classes of entities have reason to be conservative and to move slowly with new technology. Utilities have to provide electricity on demand to their customers at "least cost" while being fiscally responsible to their lenders and owners and satisfying regulatory bureaucracies. Most responsible producers want to be in business for the long run and to eventually provide financial returns to their investors and lenders. They can't afford very many hardware failures or the cost of excess warranty service or frequent liability payments. Utilities are specifically excluded from some of the government incentives. None of the incentives is directly beneficial to the producers.

Although utilities can participate in up to 50% ownership of a small power producer, there are few incentives for doing so. However, many of the utilities are helping windfarmers by providing land, power lines, and other physical and technical assistance. The utilities would like to do more and some have been lobbying for eligibility for both the tax credits and the PURPA benefits.

WECS producers can, of course, develop windfarms and serve as small power producers in addition to WECS hardware producers. At least two companies (U.S. Windpower, Inc. and Fayette Manufacturing Corporation) have limited their hardware to projects which they develop. At least another two (Hamilton Standard Division and WECS-Tech Corporation) are offering hardware to others and also developing their own windfarms. Obviously, combining the hardware production and windfarm development functions imposes additional financial and managerial demands on the producer while providing the opportunity for additional profits and risks.

The small power producer's role is to develop WECS projects with generating capacities of up to 80 MW. Depending on the deal a SPP makes with its host utility, either the SPP or the utility can operate and maintain the facility and feed the electricity into the utility's system. To take full advantage of the available tax shelter, the wind-farm must generate for at least five years. The utilities, hardware producers, and responsible SPPs want the windfarm to produce for at least 15 to 30 years without the need for major retrofit or replacement.

The driving economic motivation for small power producers is obviously the opportunity for positive cash flow and tax shelter in the first five years. With an appropriate mix of equity and debt financing (25% or less equity allows the investment to be returned by the federal tax credit in the first year of operation), financial risk to the SPP can be minimal and the opportunity for total (cash plus tax benefits) return on investment can be very high. Of course, the WECS must work and, for the long run, the selling price of the electricity and the energy out of the WECS must be high if debt is to be retired and the project remain profitable after tax credits and depreciation are exhausted.

Typically, small power producers are organized as limited partnerships to allow maximum utilization of tax credits and to minimize financial risk to the investors. The managing or general partner can be an individual, a corporation, a joint venture or a combination of those and others. The success of a windfarm will be based on the success of the general partner to manage the project technically and financially and to deal with the inherent risks. The limited partners are basically passive investors who provide equity funding and risk an amount equal to their cash investment plus any guarantees they provide for partnership debt. Limited partners can be individuals, corporations or other entities. Debt financing is usually provided by traditional sources such as banks, insurance companies, pension funds, trusts, etc. In some cases, the WECS hardware producers provide easy payment terms or help with either the equity or debt financing. Some projects are further complicated by special treatment of land, utilizing facilities or services available from the utility, or complex sale and lease-back schemes.

The small power producer doesn't have to perform all functions to develop a windfarm, but it must be able to manage those functions and serve as an entrepreneurial catalyst to bring together all of the diverse, and sometimes conflicting, elements of the project so that the WECS can be installed and operated on schedule and within budgets. The total project planning and development process is discussed in greater detail in Chapter 5, but some of the major functions include:

- Site selection;
- Project planning;
- Land acquisition;
- Project financing;
- Contract negotiation with the utility;
- Project management;
- Hardware selection and purchase;
- Hardware installation;
- Hardware operation and maintenance;
- Project disposal.

Although all of those functions are essential, the ability to negotiate a favorable contract with the utility is the one which establishes the opportunity for a successful project. Those negotiations should begin early and continue until the project is disposed of. Where possible, it should involve the utility in financing, cost sharing, facility operation, and engineering and planning of the total project. Four critical issues which must be successfully negotiated are:

- Price for the electricity, now and later;
- Duration of the contract and contingent actions;
- Interconnection requirements and costs;
- Responsibilities for operation and maintenance.

It is too early to know which of the current windfarms are successful or which small power producers will be profitable or even survive. However, experience over the next few years will lay the groundwork for the future of windfarms by small power producers, and government and regulatory agency attitudes toward them. The remainder of this chapter attempts to identify windfarms planned, under construction or in full operation.

Government-Sponsored Windfarms

The U.S. government sponsored two experimental windfarms to help answer technical questions about clusters of large WECS interconnected with utility grids. The first of the two is hosted by the Bonneville Power Administration (BPA) and the Klickitat County Public Utility District in the Goodnoe Hills above the Columbia River Gorge in southern Washington near Goldendale. The second is located at a U.S. Department of Interior site near their Bureau of Reclamation (BuRec) Colorado River Storage Project near Medicine Bow, Wyoming.

Three DOE/NASA 2.5 MW MOD-2 experimental HAWTs were built for the $40 million BPA project by Boeing Engineering and Construction Company starting in August, 1977. The site was chosen based on three years of meteorological testing conducted by Oregon State University for BPA. By mid-1980 BPA had a small substation in place to raise the anticipated WECS 12.5 kV output to the 69 kV needed in the Northwest electric grid. The first HAWT nacelle was installed on its 193 ft high tower in September, 1980 and the first 300 ft long rotor was installed the following month. By May, 1981 the three unit 7.5 MW capacity mini-windfarm was standing and ready for start-up, debugging, and testing. (See Figure 3-19.)

Unfortunately, on June 8, during routine testing, one of the units went into an overspeed condition (accelerating from its planned speed of 17.5 rpm to 28 rpm in less than 30 seconds) which burned out the electrical system and took the unit out of service. The other two units, although not damaged, were shut down until the cause of the failure could be determined and corrective action taken. After modifications, the two undamaged units were turned on in the fall of 1981 and the repaired unit was put back into service in the spring of 1982.

With the delay in getting the large HAWTs working, the windfarm R&D was also delayed. The government hopes to learn more about such things as spacing of large machines in clusters, and the quality of electricity output as windspeeds vary and as each machine cuts in and shuts down under different wind conditions. The three HAWTs were purposely positioned at the corners of an irregular triangle whose sides are five, seven, and ten blade diameters (1500 ft, 2100 ft, and 3000 ft) long. Researchers from Oregon State and Battelle Pacific Northwest Laboratory will be testing the effects of the machines on one another at different spacings and under different wind

Figure 3-19. BPA's 7.5 MW experimental windfarm: three DOE/NASA MOD-2s installed in Goldendale, Washington, supplied by Boeing Engineering and Construction Company.

characteristics to determine the extent of air turbulence and wind energy reduction in the wake of each machine. Those studies will help determine how close rotors can be spaced and, in turn, the energy density potential per acre of land.

BPA has estimated that the area around the Goodnoe Hills could support a 100 MW windfarm and that its output could be comfortably absorbed into BPA's Northwest power grid.

Over in Wyoming, delays have plagued the first phase of BuRec's ambitious 100 MW windfarm near Medicine Bow. Two system verification units (SVU) were expected to be operational in 1981 but were delayed until mid 1982, at the earliest, by lack of funding and other problems. One SVU, a DOE/NASA 2.5 MW MOD-2 HAWT has been up since December, 1981 but neither DOE or BuRec was budgeted funds to operate or test it. As a result, the prototype sat until DOI could redirect some funds and get start-up and testing activities underway in May, 1982. The second SVU, Hamilton Standard's first 4 MW capacity WTS-4 HAWT, is under construction but not expected to be ready for testing until the fourth quarter of 1982 or early 1983. BuRec funding of the WTS-4 unit is available. (See Figure 3-20.)

The Medicine Bow project has been under study since 1976 as BuRec planned a series of windfarms at windy sites accessible to their existing hydroelectric system. BuRec proposed integrating giant windfarms with the existing dams, reservoirs and powerplants in the Colorado River Storage Project. It hoped to conserve water and make its existing facilities more efficient by storing water behind such dams as Flaming Gorge and Glen Canyon when the wind blows, and generating hydroelectric power when the wind stops. Extensive BuRec studies showed that the idea was sound and that winds available near their facilities could provide as much as 10% of the power needs of the Rocky Mountain area by the year 2000. The first two WECS, as well as the first 100 MW windfarm, were intended to demonstrate feasibility of the concept and allow BuRec to move toward larger windfarms with confidence in both WECS hardware and the interrelationships between WECS and hydro generating systems.

The Medicine Bow 100 MW windfarm is projected to deliver 300 to 350 million kWh per year to be marketed by the government's Western Area Power Administration (WAPA). Costs for 25 WTS-4s and 40 MOD-2s were estimated in 1980 to be $185 million and $275 million, respectively, and BuRec asked congress to budget $189 million for the project with the objective of securing competitive bids for the actual project. Because over two-thirds of the electricity generated is expected to coincide with peak demand, a major portion of it in the winter, integration of WECS with BuRec's hydropower is especially valuable.

Figure 3-20. BuRec's system verification units: DOE/NASA's 2.5 MW capacity MOD-2 and Hamilton Standard's 4 MW Capacity WTS-4 MOD-2 shown.

Although prospects of congress funding the entire Medicine Bow windfarm are not encouraging, BuRec's planning and prototyping work has been valuable in identifying and overcoming many issues associated with large scale development of windfarms. If congress decides not to invest in WECS directly, there should be an opportunity for small power producers to use the impressive BuRec data base and voluminous studies to move quickly into construction of numerous windfarms which can generate the electricity to satisfy the BuRec and WAPA needs.

Privately Sponsored Windfarms

In the private sector, the leading small power producer, in every way except getting WECS up and operating, is San Francisco based Windfarms, Ltd. They have over a billion dollars worth of agreements with utilities.

Windfarms was started in 1978 by entrepreneur Wayne K. Van Dyck and by 1981 it had blue chip investors, a quality staff of forty people, world class advisors and consultants, and windfarm projects on the drawing boards for Hawaiian Electric Company (Honolulu), Hawaiian Electric and Light Company (Hilo), Maui Electric Company (Kahului), Pacific Gas and Electric Company (San Francisco), and the California Department of Water Resources (Sacramento). They also were pursuing projects in southern California (over 10,000 acres were acquired or leased which were believed suitable for 400 MW of WECS power which could be sold to Southern California Edison Company) and Oregon (three sites believed capable of supporting 200 MW of WECS power for sale to private Oregon utilities and Bonneville Power Administration). Obviously, all was going well.

Less than a year later, in March, 1982, Windfarms, Ltd. announced it was sharply cutting back its operations, consolidating its activities, reducing its staff from 30 to 19 employees, and deferring for at least six months all work on its 350 MW project with PG&E and CalWater. Things were not going so well.

Oil prices were down and so were projected windfarm revenues which used oil as the index for inflation adjustments in prices utilities would pay for WECS electricity. High interest rates were still in force long after they were projected to fall. PURPA was losing early skirmishes in the courts. The Reagan administration had not only abandoned WECS R&D, it was less than enthusiastic about tax credits for alternative energy. And the recession started to hit bottom.

Wayne Van Dyck had anticipated many of the problems and had worked to give Windfarms, Ltd. what he called "staying power." His own initial investment was followed by a capital infusion of $1 million by two San Francisco area venture capitalists (Brae Partners and Prim Investments) in August, 1979. In early 1981 another $16 million was invested in the company by Standard Oil Company of California ($10.7 million for 40% equity) and Sigma Resources Group, Inc., an affiliate of New York-based Madison Fund ($5.3 million for 20% equity).

Van Dyck's business strategy included working with the most reputable "associates" that could be found to build credibility in an emerging industry with no track record. The Windfarms list of engineering, finance, banking and legal consultants read like a who's who in the professions. His list of preferred WECS producers read like a who's who of American industry.

So why weren't there some Windfarms, Ltd. WECS in operation in 1982? Although Windfarms has been careful not to point the finger of blame, none of the industrial giants working on the large WECS had yet come through with commercial hardware. A deal with Hamilton Standard for 20 of their 4 MW HAWTs for the Hawaiian Electric Company project fell through for a variety of reasons, but Hamilton Standard is still working to get its prototype up in 1982. A deal with Alcoa for six to 22 of their 500 kW VAWTs for the smaller Hawaiian projects fell through when one of Alcoa's prototypes self-destructed in the San Gorgonio Pass. Westinghouse, Boeing and General Electric weren't ready to take orders for commercial versions of their designs for DOE/NASA. With all the other problems, it is not surprising that Windfarms, Ltd. had to acknowledge reality and slow down.

Although not yet under construction, two of the Windfarms pioneering projects deserve attention. In July, 1979 Hawaiian Electric Company (HECO) signed a letter of intent with Windfarms for an 80 MW project. After nine months of negotiation, a power purchase contract was signed under which HECO agreed to purchase the power from the project for 25 years and to build a 19 mile transmission line. A lease on 2300 acres at Kahuku was then worked out with the Campbell Estate. Simultaneously, an arrangement with the U.S. Army to share the site for training maneuvers was reached. In the summer of 1980, the Hawaiian legislature passed legislation which cleared the way for the project.

The Kahuku site is on the north shore of the Island of Oahu near the site where the DOE/NASA 200 kW MOD-0A has been gathering valuable operating experience. The tradewinds provide an excellent wind resource. Fuel oil is currently used to generate 97% of Oahu's electricity, providing a high competitive cost.

Although WECS hardware, project schedules, and costs are not yet known, the 80 MW windfarm is expected to supply almost 9% of Oahu's annual electricity needs, serving almost 100,000 people and saving almost a million barrels of imported oil each year.

The northern California windfarm with Pacific Gas and Electric and California Department of Water Resources, although delayed, is important because of its size (the largest wind project yet announced, at 350 MW) and the interrelationships of one of the largest investor owned utilities (PG&E) and one of the largest state agencies (Cal-Water). The project involves Windfarms, Ltd. as the builder, owner and operator of the project, PG&E as the purchaser and marketer of the electricity generated, and CalWater as the purchaser of off-peak energy to help meet its pumping requirements for the California State Water Project. The site is in Solano County, approximately 30 miles northeast of San Francisco and 50 miles west of Sacramento in a rough triangle bounded by Vallejo, Benicia, and Fairfield and Interstate Routes 80, 680, and 780. The excellent wind conditions at the site indicate generation of almost a billion kWh and savings of 1.6 million barrels of oil annually when the 350 MW windfarm is completed.

When the formal electricity purchase contract was signed on January 7, 1982, PG&E Vice-President for Planning and Research Nolan H. Daines hailed it as a milestone. Daines indicated the contract "is the largest of its kind ever negotiated by a public utility. It sets the stage for the largest windfarm yet planned in this country. And it further confirms PG&E's confidence in the future of renewable resources." For its role in the project, Department of Water Resources Director Ronald B. Robie saw the contract moving the state water project closer to its goal of supplying 70% of its energy needs from renewable resources. Robie said, "The Solano windfarm will move us from research to operational status in wind energy. We are proud to be a part of this undertaking."

The first private windfarm to generate electricity was turned on in southern New Hampshire on the last day of 1980. The 600 kW array of twenty 30 kW HAWTs was built by U.S. Windpower, Inc. of Burlington, Massachusetts and interconnected with Public Service Company of New Hampshire lines on a windy ridge on Crotched Mountain (see Figure 3-21) and became the first small power producer to become a "qualified facility" under rules of PURPA. By feeding a small amount of electricity into the utility lines on December 31, it also became the only windfarm to take advantage of tax credits in 1980.

Unlike most other small power producers, U.S. Windpower de-

Figure 3-21. The first private windfarm, built by U.S. Windpower, Inc. on Crotched Mountain, New Hampshire.

signed and built its own wind turbines. The original three-bladed 40 ft diameter downwind HAWTs were based on a University of Massachusetts "wind furnace" prototype. U.S. Windpower has a 20 year contract to sell electricity to Public Service and a separate contract to lease the land from the Crotched Mountain Foundation and supply their rehabilitation center with low-cost electricity.

The U.S. Windpower project in New Hampshire, just two hours from its headquarters, served as a prototype for what they hope will be a series of projects in the Pacific Northwest, New England, Wyoming, Montana, Hawaii, and California. Their first large windfarm is currently being built in three phases in the Altamont Pass, 50 miles east of San Francisco, in cooperation with Pacific Gas and Electric.

For the California project, U.S. Windpower uprated their HAWT to 50 kW by increasing the rotor diameter to 56 ft. During Phase I, 100 machines were installed on the Walker Ranch in early 1982. (See Figure 3-22.) Phase II is being built on an adjacent site and will add another 100 units in mid 1982 for a total of 10 MW. Equity and debt funding for Phase II was raised by a $11 million offering by stockbroker Merrill Lynch in March, 1982. The resulting limited

Figure 3-22. U.S. Windpower's California windfarm. 100 50 kW HAWTs (5 MW) are shown as Phase I. Another 500 Units are planned nearby.

partnership was designated Windpower Partners, Ltd. The third phase is expected to add another 400 HAWTs, on a separate site, and bring the total to 600 machines. When completed the 30 MW windfarm is expected to generate approximately 90 million kWh per year for sale to PG&E.

A companion project in Altamont Pass is being constructed by Great Falls Wind Energy Corporation, associated with Fayette Manufacturing Corporation of Clearfield, Pennsylvania. 34 of Fayette's three-blade, 30 ft diameter HAWTs were installed on 80 ft towers on the Joseph J. Jess ranch by the spring of 1982. (See Figure 3-23.) The WECS have 75 hp induction generators which are expected to generate peak power of 56 kW in very high winds. A total of 300 machines are planned for the Jess ranch for a windfarm capable of over 15 MW of power interconnected with PG&E's lines.

Three windfarms were being built in 1981 in Boulevard, just north of the Mexican border about 45 miles east of San Diego, for sale of electricity to San Diego Gas and Electric Company. Mehrkam Energy Development Company (MEDC) of Hamburg, PA was sup-

Figure 3-23. Fayette's California windfarm. 34 50 kW HAWTs (1.7 MW) are shown installed in April, 1982.

plying and installing 40 kW, 100 kW, and 225 kW capacity HAWTs in hopes of getting machines turned on by December 31. However, tragedy struck on December 30 when MEDC President Terry Mehrkam was killed while working with one of his 40 kW units during a storm which caused several machines to run out of control and lose their blades. At this writing, the future of the three windfarms is uncertain.

The Southern California Edison Company (SCE) service area had 24 windfarms in varying stages of planning and development on May 21, 1982. If all of the projects are completed and brought on line, 378–490 MW of power will be added. SCE encouraged windfarm development by requesting proposals for what they designated "wind parks" in 1980. SCE also assigned an experienced engineering manager, Bill Emrich, as project manager for wind commercialization to help get the "wind parks" into service. By the spring of 1982, seven parks (76.8–91.8 MW capacity, depending on options) were under contract, four (53.9–56.9 MW) had negotiated letters of intent, and another thirteen (247–344 MW) were at the stage of serious negotiations.

Most of the wind parks are planned for the windy San Gorgonio Pass area of Riverside County near where SCE has its own Wind Energy Center. However, construction in that area was delayed while the county and the Bureau of Land Management studied environmental impacts and developed a land use plan. That action is complete and windfarms are expected to be under construction in 1982.

For San Gorgonio, Hamilton Standard Division, United Technologies Corporation, Windsor Locks, Connecticut is planning a 20 MW farm with five of its WTS-4s; WECS-Tech Corporation, Gardena, California is planning a 5.5 MW project with 55 of its own 100 kW HAWTs; Pan Aero Corporation, Golden, Colorado is working on 30 MW; a joint venture of First National Capital, Birmingham, Michigan and Manley, Bennett, McDonald and Company is planning a windfarm with sixty of Westinghouse's WWG-0500 HAWTs; and Ventus Energy Corporation, Covina, California is proceeding with a 30 MW farm in cooperation with Energy Unlimited, Villanova, Pennsylvania.

The first wind parks actually up for SCE are in the windy Tehachapi Mountains area. (See Figure 3-24.) Zond Systems, Inc., Santa Ynez, California built the first phase of its Victory Garden project utilizing 15 of Wind Power Systems' 40 kW Storm Master Model 12 (see Figure 3-25) and later ordered two of the DAF-Indal 500 kW VAWTs.

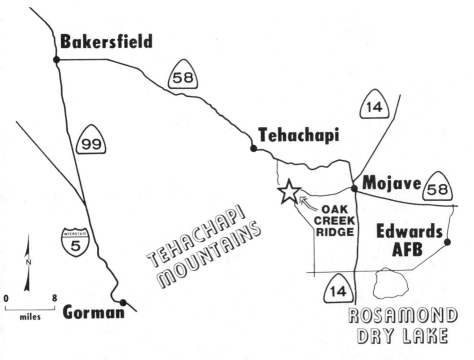

Figure 3-24. The Tehachapi Wind Resource Area, site of several windfarms for Southern California Edison.

Figure 3-25. Zond Systems Victory Garden windfarm; Wind Power Systems' 27/40 kW HAWTs shown.

Before completion, Victory Garden is expected to grow to a total capacity of 30 MW.

Ridgeline Windfarm, Beverly Hills, California, is developing its 5 MW windfarm on Oak Creek Ridge in the Tehachapis. Pacific Wind and Solar has installed ten of Energy Sciences, Inc.'s 50 kW ESI-54 HAWTs at that site. (See Figure 3-26.)

The Beckett-Cummings-Soule partnership, doing business at Oakcreek Energy Systems, is proceeding with a 5 MW windfarm in Tehachapi which they hope will grow to 20 MW. Initially, WECS hardware is being supplied by Carter Wind Energy (their 25 kW HAWT) but a variety of newer designs are being explored by the partnership.

American Wind Energy Systems, Bakersfield, California, is planning another 4 MW windfarm in the Tehachapis, but details are not yet available.

Outside of California, windfarm activity is more sparse. In Texas, a much publicized windfarm in Dalhart got off the ground in the late '70s but ran into technical problems and was never completed. Therefore, the first Texas windfarm didn't come on line until late 1981. Osborne Solar, Elgin, Texas, installed five Carter Wind Systems 25 kW HAWTs at a site north of Pampa, Texas (see Figure 3-27) and tied four of them (the other serves a large home) into the Southwestern Public Service Company grid in October as a mini-windfarm. Another four Carter 25 kW units were installed in Livingston, Montana in 1981 as the nation's first municipal owned windfarm. The 100 kW Livingston project is interconnected with Montana Power Company's grid.

In Vermont, a partnership called 1981 Wind Investors was created to build and own a one machine, 200 kW windfarm on Little Equinox Mountain. A WTG Energy Systems MP-200 HAWT was constructed on land owned by the Carthusian Foundation and its first electricity was sold to Central Vermont Public Service Company on December 31, 1981 to qualify as a small power producer and 1981 tax credits. (See Figure 3-28.)

Several windfarms are planned for Hawaii, but the first to get a machine up is Carl Huntsinger's Molokai Energy Corporation. Molokai Energy earlier installed a Mehrkam 40 kW HAWT on windswept Molokai Ranch and followed that up with the first of five ESI 50 kW units in 1981. Growth of the windfarm is inhibited by the fact

Figure 3-26.   Ridgeline Windfarm; Energy Sciences' 50 kW HAWTs utilized.

Figure 3-27. The first Texas windfarm; Carter 25 kW HAWTs by Osborne Solar near Pampa.

Figure 3-28. The one machine Vermont windfarm, a WTG Energy Systems MP-200 HAWT installed on Little Equinox Mountain.

that the island's total power demand from its utility, Molokai Electric, is less than 10 MW.

The concept of generating bulk power with windfarms, either owned by utilities or by small power producers, has been proven. Many other projects are in the planning stage and will, no doubt, be announced by the time this book is being read.

4
Dispersed Applications: Investing in the Future

Although the major WECS projects are currently being developed by special purpose windfarmers, there are many opportunites for using one or more WECS "on-site" in dispersed applications in windy parts of the U.S. If you own or have control of an unobstructed, windswept piece of ground, you have a good start toward a local money-making WECS installation or project.

One way to use that land is to be a windfarmer and fully develop the WECS project yourself. Another way is to invest the land as part of a partnership or cooperative venture and let an experienced developer put the project together and then provide you with equity and cash flow as your winds are harvested in the form of electricity sold to your local utility.

The best opportunities are in non-urban areas where the land is unobstructed and buildable. All WECS need access to the wind. If more than one wind turbine is installed, each must be installed in a location that does not block another's wind and is not obstructed itself. Depending on the specific wind turbine and wind conditions, required spacing will be 6–15 times the WECS rotor diameter in line with the wind. If winds are near unidirectional, the perpendicular spacing can be in the range of 2–5 rotor diameters.

The buildable part of the land requirement relates to physical site conditions, to practical considerations, and to regulatory agency requirements. Alternative uses of that available land also should be considered. If your site is inaccessible, is a swamp, or is vertical, it will be expensive to build on and probably not be a good one for WECS. If it is relatively flat and looks like it will support WECS foundations and anchors, investigate further if it is near a road and a power line.

From a practical standpoint, delivery trucks, erection cranes, and workers have to get to the site, and the installed WECS has to have a way to get its electricity into the power line or to electrical loads for direct utilization. If the power is for the utility, a substation or transformer, or some other means of tapping into their line in a safe way at the proper voltage, is essential. The same needs are there if you're tapping into your own electrical system. Will your lines handle the new power? Do you have three-phase circuits if the WECS is going to need or generate three-phase power?

If all else looks encouraging, make sure that the regulatory agencies who might claim jurisdiction won't find some reason to stop your project. If your property is zoned, are WECS allowed or disallowed? Will design and construction be subject to a local building code and require a permit? Can your local utility cooperate with you or do they have to go back to a regulatory agency for approval? Is the utility itself capable of dealing with innovation, or does it have built-in regulations or inflexibility which will stop, delay or add cost to the WECS project? There is one set of rules for WECS installations with less than 100 kW generating capacity, another for up to 30 MW, a third for up to 80 MW and many others if you install over 80 MW and, essentially, become a utility yourself.

A more detailed discussion of the WECS development process is offered in the final chapter of this book, but if you don't have good winds and land, you don't have to bother with the rest.

This chapter attempts to identify some dispersed applications for WECS of substance for both potential private and public business operations. Those whose business is primarily the generation, transmission and distribution of electricity (the electrical utilities) are covered in the previous chapter.

The dispersed opportunities are broadly classified and covered as:

- Agribusiness operations;
- Irrigation and water pumping;
- Business, industry, and the major processes;
- Governments and the institutions.

AGRIBUSINESS OPERATIONS

Although America's farms and rural homes are the most experienced users of windpumpers and windchargers (an estimated six million small machines have been installed since the American style multivane windpumpers were developed in 1854 by Daniel Halladay and John Burnham), utilization of modern 60 Hz, AC farm-compatible WECS presents new opportunities and potential problems.

DOE's Rocky Flats, Colorado SWECS Research Center has been working since the mid-'70s to further develop small WECS specifically for farm and rural home use and to both identify and reduce problems and barriers to widespread installation. Today, the market for the established very small SWECS (such as by Aeropower, Bergey, Dunlite, Enertech, Jacobs, Kedco, Millville, North Wind, Sencenbaugh, Tumac, Winco, Wind Power Systems, and Windworks) is predominantly in rural America.

This book acknowledges those small machines (and the author encourages purchase and installation of them for known windy sites without spending a lot of money for studies), but is directed toward more substantial energy applications, those which can use 20 kW of electric power or more.

Although U.S. Department of Agriculture (USDA) statistics show that over half of on-farm energy is used for transportation and field work, WECS can't help much with those moving applications. Of the 45 percent consumed as stationary energy on the farm, USDA lists the following as the major users:

- 39 percent for irrigation;
- 29 percent for heating and cooling buildings;
- 22 percent for product processing;
- 6 percent for farmstead electrical power;
- 4 percent for all other work.

The USDA has been encouraging responsible utilization of WECS since the mid-'70s and has funded active applications-oriented R&D with emphasis on economic viability and reliability. The early USDA program, with help from DOE, was managed by Lou Liljedahl of the

Agricultural Research Service (ARS) in cooperation with Bob Meroney of Colorado State University representing the Cooperative State Research Service (CSRS). It concentrated on specific dedicated applications of WECS on the farm and executed several projects. The primary thrust was in three areas:

- Product processing and storage, managed by USDA-ARS in Beltsville, Maryland by Herschel Clueter;
- Building heating, managed by USDA-ARS in Ames, Iowa (Iowa State University) by Leo Soderholm;
- Irrigation, managed by USDA-ARS at the Southwestern Great Plains Research Center in Bushland, Texas by Nolan Clark.

Virginia Polytechnic Institute demonstrated an apple cooling and storage scheme and Kaman Sciences demonstrated milk cooling under the WECS product processing and storage category.

Cornell University demonstrated water heating in a dairy milking center and ARS undertook in-house demonstration programs for milk cooling and water heating in Beltsville, and for farmhouse heating in Ames.

Major in-house irrigation applications R&D efforts were undertaken in Kansas (shallow lift irrigation) and at Bushland (deepwell irrigation).

Although the dedicated applications were technically successful, they led to the conclusion that WECS should be utilized as general supplements for other power and fuel sources on a broader, total farm basis. The broader approach reduces the problems of trying to match availability of the intermittent fuel (the wind) with the intermittently used on-farm processes. It also removes the need for dedicated on-site storage (such as batteries) by integrating WECS with the local utility system.

By integrating WECS with the utility's electrical system, electricity is available on demand and can be purchased when the wind is not available. Conversely, when the wind is powering the WECS at a time when the electricity is not needed on the farm, it can be sold to the utility. The interconnection with the utility lines allows installation of the most economically attractive size or number of WECS without having to match maximum WECS power with power load demands.

Within the general classification of agribusiness operations, the best use of WECS power is to supplement the local utility. Although most agribusiness operations are in rural locations, few are at locations remote from the power company. If an application is remote, the best WECS application is one integrated with an on-site power plant (such as an internal combustion generator) rather than relying on dedicated storage systems or attempting to combine WECS with other intermittent systems (such as solar thermal or solar photovoltaics). In this combination, the value of the WECS is directly relative to the value of fuel saved in the on-site power plant.

Where the utility power is available, the value of the WECS will be either the value of the electricity not purchased, the value of the electricity sold to the utility, or, in most cases, a combination of the two.

Unless the utility pays more for the electricity when it buys it back than it charges when it provides it, it is common to size the WECS so that maximum power expected from the WECS equals the minimum power (base load) utilized in operations. However, if the utility's buy-back rate is high enough to allow a profit on the sale of the wind powered electricity, the size of the WECS will be limited only by available land and the capacity of the utility to accept electricity before its value (and the buy-back rate) goes down. These considerations are covered in more detail in the final chapter and are heavily dependent on tax considerations, available incentives, the utility's ability and willingness to cooperate, and your financial resources and objectives.

If you expect to utilize all of the wind powered electricity in your operations, choosing the right size WECS is important. It may turn out that a very small (less than 20 kW capacity) WECS is all you can justify. On the other hand, if operations are energy intensive (such as in dairy farming, poultry confinement, food freezing or cooling, high-temperature crop drying, livestock confinement and processing, grain elevators, and other product processing and storage) WECS with generators well in excess of 100 kW may be justified. The probable best size WECS for agribusiness applications will be in the 20–200 kW size range.

In addition to the impressive USDA applications R&D efforts (and hardware and development efforts by DOE's Rocky Flats SWECS

Center and Sandia Laboratories VAWT program), important agribusiness oriented WECS programs were undertaken in the '70s by both Clarkson College of Technology and the Alternative Energy Institute at West Texas State University.

Clarkson, led by Edward Kear, initiated two WECS projects in the Potsdam, New York area, relative to dairy farms. The first resulted in a 15 kW HAWT (the pioneering Grumman Windstream 25, no longer offered) being installed in 1977 on a working dairy farm operated by the David J. McKnight family in Hopkinton, New York in cooperation with New York State Energy Research and Development Authority and Niagara Mohawk Power Company. A year later, Clarkson headed a broader development team which installed a silo-mounted Darrieus VAWT at a Clarkson-owned site near the Potsdam airport. Dr. Kear involved Alcoa, Madison Silo (Chromalloy Farm Systems), Reliance Electric, Agway (the Northeast's leading agribusiness cooperative) and others in addition to Niagara Mohawk. The innovative silo-mounted WECS hardware project was based on the concept of installing the WECS where its output could be utilized almost directly, using the always available dairy farm silo as a "free" structural tower to get the WECS high in the wind and free of obstructions, and utilizing the conical shape of the silo roof to accelerate and focus the wind on the WECS. (See Figure 4-1.)

The Alternative Energy Institute (AEI) was created in the fall of 1978 by West Texas State University (in Canyon) and is the outgrowth of efforts which began in 1970 by Vaughn Nelson (WTSU) and Earl Gilmore (Amarillo College) in the area of wind energy. The AEI has worked closely with USDA's nearby Bushland WECS irrigation program and also has been a national leader in farm related WECS information dissemination and development efforts. The AEI has advocated cooperative effort between government agencies, utilities, WECS providers and WECS users, and has been in the forefront of organizing such cooperation. (See Figure 4-2.)

Dr. Nelson also served as a leader in establishing the American Wind Energy Association and serving on its early boards of directors. He believes "There is a large market for wind turbines in the rural sector as WECS from 10 kW to 100 kW can produce power for stationary uses. Farmers and associated agricultural industries will become interested in windfarms. This means that in the future the farmer will

Figure 4-1. Clarkson College silo-mounted VAWT, a cooperative venture by Agway, Niagara Mohawk, Alcoa, and others in 1978.

farm the wind much as they now farm the sun. In addition, through use of storage in the form of chemicals, gas, and flywheels, WECS will be used as power sources for nonstationary uses of energy."

IRRIGATION AND WATER PUMPING

Irrigation now consumes approximately 11 percent of all energy used in agriculture (the equivalent of about 87 billion kWh in 1978, according to USDA). Between 12 and 15 percent (millions of acres) of all cropland is irrigated to yield almost 30 percent of all agricultural products. Almost 500 thousand on-farm plants pump and deliver irrigation water, of which almost half are in the windswept Great

Figure 4-2. Alternative Energy Institute test program, a 32 ft diameter DOE/UTRC 15 kW prototype installed and tested by Vaughn Nelson's AEI on a Phillips Petroleum oil well lease near Borger, Texas.

Plains/Lower Rocky Mountain states of Colorado, Kansas, Nebraska, New Mexico, Oklahoma, and Texas.

Irrigation requires energy both for lifting the water to ground level and for distribution of the water on the field. Lifting the water may require pumping from surface sources, such as rivers, shallow wells, lakes, farm reservoirs, or tail-water pits, which may vary from a few feet to as deep as 75 ft, or pumping from a well in an aquifer which may require lifting from as deep as 750 ft. The energy required for lifting is, of course, directly proportional to the lift. Distribution is by center pivot and fixed sprinkler installations, and gravity systems such as gated pipe and header-ditch/siphon systems.

The USDA's Conservation and Production Research Laboratory at the Southwestern Great Plains Research Center in windy Bushland, Texas, has managed an impressive WECS irrigation program since the mid-'70s. In close cooperation with the AEI, DOE's Sandia National Laboratories, and several U.S. and Canadian WECS producers and consultants, the USDA program has been productive in assisting with WECS hardware development and testing, applications research and demonstration, and practical guidance toward economically attractive WECS utilization.

In response to rapidly escalating conventional fuel costs and, in some cases, the threat of shortages, USDA moved quickly into applications research and prototype testing to get hands-on experience with WECS in irrigation. USDA's Agriculture Research Service (ARS) in Kansas, lead by Lawrence Hagen, concentrated on shallow-lift water pumping which led to installation and testing in Garden City of a 30 ft × 20 ft two-bladed Darrieus VAWT (supplied by DAF-Indal) which pumps water from a tail-water pit to the fields for direct irrigation and to a reservoir for storage and later gravity-flow irrigation. At Bushland, Nolan Clark's team proceeded with an ambitious program for deep-well pumping and water distribution. By 1981, USDA and AEI had seven WECS directly and indirectly pumping water. The seven WECS utilized were:

- 18 ft × 15 ft two bladed Darrieus VAWT (4 kW) produced by DAF-Indal (see Figure 4-3);
- 55 ft × 37 ft two bladed Darrieus VAWT (40 kW) produced by DAF-Indal (see Figure 4-4);
- 83 ft × 55 ft two-bladed Darrieus VAWT (100 kW) produced by Alcoa through DOE/Sandia Laboratories (see Figure 4-5);
- 32 ft diameter, two-blade, downwind HAWT (25 kW) produced by Jay Carter Enterprises (three units; see Figure 4-6);
- 40 ft diameter, three-blade, downwind HAWT (25 kW) produced by Wind Engineering Corp. (see Figure 4-7).

Because deep wells require substantial power, the major USDA effort has been with WECS for use in three potential modes of operation:

- Wind assisted internal-combustion engines (using diesel oil, natural gas, LPG, and gasoline as fuels; see Figure 4-8);
- Wind assisted electric pumps (with and without sale of excess electricity generation; see Figure 4-9);
- Stand-alone systems (with gravity-flow reservoirs).

Figure 4-3. USDA's 4 kW VAWT, research machine supplied by DAF-Indal for irrigation project.

The wind assisted combinations have advantages:

- Water is available when needed;
- Constant pump rpm is maintained for efficiency;
- In-place systems are easily retrofitted;
- Constant water flow allows good irrigation efficiency;

and disadvantages:

- They require two power sources;
- Utility demand charges usually aren't reduced;
- If mechanically linked, WECS operate only when the load is operated;
- They usually are more expensive than stand-alone systems.

In the Great Plains region alone, "Successful windpowered irrigation pumping would be applicable to approximately 200 thousand

Figure 4-4. USDA's 40 kW VAWT, one of DAF-Indal's early VAWTs used by USDA for deepwell irrigation research.

irrigation wells," according to USDA's Dr. Clark. Because more than 60 percent of the pumped irrigation water is powered by fuels in internal combustion machines, rather than by electricity, the USDA program has attempted to integrate both electricity generating and direct shaft power WECS into irrigation systems.

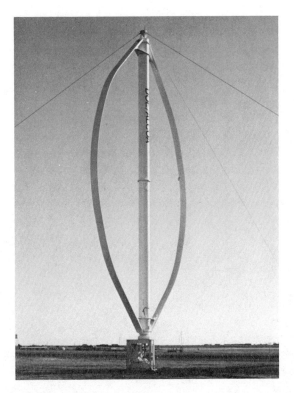

Figure 4-5. The DOE 100 kW VAWT at Bushland, built by Alcoa for DOE/Sandia in close cooperation with USDA's wind research team, this prototype is one of the U.S. government's most successful wind ventures.

Although natural gas was long the most economical fuel for irrigation pumps, rapid escalation of prices (as well as non-availability in many regions) has increased the use of electricity and diesel fuel in the '80s. Therefore, the economics of wind have been explored in competition with all of the fuels, including electricity.

Two major 1980 irrigation studies by Southwest Research and Development Company (Las Cruces, New Mexico) for USDA combined data on type and size of pumps, cost and availability of fuels for those

Figure 4-6. The Carter 25 kW prototype at Bushland. This machine served as the test bed for refinements and retrofits which led to Carter Wind Systems commercial Model 25.

pumps, and current and projected economics of WECS in various modes of operation in the following irrigation regions of the U.S.:

A. Kansas, Oklahoma, High Plains of Texas, High Plains of New Mexico, and Eastern Colorado;
B. Nebraska, South Dakota, and North Dakota;
C. South Texas (Edwards Plateau) and the Pecos Valley area of New Mexico;
D. Southern Arizona and Southern California (primarily the Imperial Valley);

Figure 4-7. The Wind Engineering 25 kW prototype at Bushland. R&D Work on the Wingen 25 led to Coy Harris' commercial model 25–42.

   E. Idaho, Oregon, and Washington (Snake and Columbia River basins);

   F. Midwestern U.S.—Illinois and Indiana;

   G. Southeastern U.S.—Florida and Georgia.

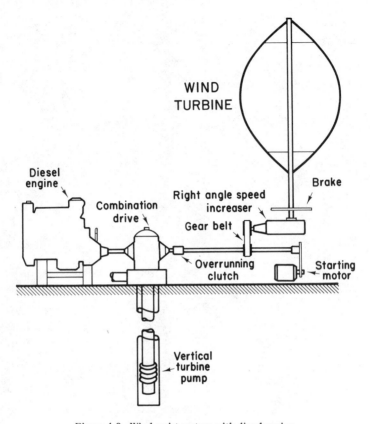

Figure 4-8. Wind-assist system with diesel engine.

Those studies concluded that WECS installed in both "wind-assist" and "stand-alone" modes can be cost-effective options in parts of regions A and B, especially in competition with diesel and LPG fuels. Although there is good correlation in those regions between wind intensity and the irrigation season (March through October in Region A and April through September in Region B), economics are improved if excess WECS-generated electricity is also utilized in other on-farm operations or sold at a reasonable price to the utility.

As always, availability of strong winds is the most important consideration in determining feasibility of a WECS for irrigation pumping. In parts of four states in USDA's Region A, available wind power

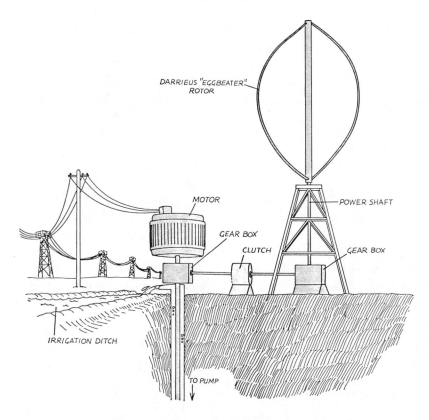

DARRIEUS "EGGBEATER" ROTOR

MOTOR

POWER SHAFT

GEAR BOX

CLUTCH

GEAR BOX

IRRIGATION DITCH

TO PUMP

Figure 4-9. Wind-assisted electric pump concept.

averages over 400 Watts per vertical square metre of WECS rotor area. At sites in that area, wind assisted pumping is believed to be cost-competitive with both diesel and LPG fueled engines today and is expected to become competitive with natural gas as the price of that fuel rises with deregulation. Gasoline is not widely utilized in the windy region. Competition with electricity is a function of site-specific availability and the price, as well as utility buy-back rates and policies.

Obviously, without favorable selling prices for excess electricity, the best combination will utilize a WECS smaller than the pump's needs so that the WECS will operate in the fuel saving mode most of the time. If the electricity can't be used on-farm during non-irrigation periods, low selling prices will reduce economic feasibility. If high buy-back rates are in effect, the irrigator can use as large a WECS as the power lines can handle and can act as a "windfarmer" (small power producer) as well as a pumper of water.

In the total U.S.A., approximately 60 percent of the 475 thousand irrigation pumps are smaller than 50 hp (37.3 kW) and a little over nine percent are larger than 100 hp (74.6 kW). However, in the windy region, the percentage shifts toward larger pumps. Of the estimated 30 thousand pumps in the windy portion of Region A, over 60 percent are larger than 50 hp (37.3 kW), and almost 18 percent are larger than 100 hp (74.6 kW). Information excerpted from the Southwest Research and Development Company 1980 study of irrigation pumping plants for USDA summarizes pumping utilized in the best wind areas of Region A. (See Table 4-1.)

If you have land in Region A, or other parts of the country with equivalent winds, exploration of the feasibility of installing one or more substantial WECS to supplement conventional fuels or purchased electricity appears to be warranted.

Table 4-1. Irrigation Water Pumping and Distribution, USDA High Wind (Over 400 W/sq. m) Portion of Region A.

NUMBER OF PUMPS	COLORADO	KANSAS	OKLAHOMA	TEXAS	TOTAL REGION
For ground water	780	19,984	2,741	6,400	29,905
For surface water	–	647	20	–	667
TOTAL	780	20,631	2,761	6,400	30,572
< 18.7 kW	–	4,763	46	142	4,951
18.7–37.3 kW	117	5,930	345	932	7,324
37.3–56.0 kW	328	4,695	659	1,350	7,032
56.0–74.6 kW	150	3,110	852	1,739	5,851
74.6–111.9 kW	181	1,729	672	1,678	4,260
> 111.9 kW	4	404	187	559	1,154
Energy source:					
electricity	148	5,158	469	256	6,031
diesel oil	16	3,714	138	384	4,252
natural gas	616	9,903	1,685	5,696	17,900
LPG	–	1,650	414	64	2,128
gasoline	–	206	55	–	261

BUSINESS, INDUSTRY, AND THE MAJOR PROCESSES

Economic feasibility of WECS of substance usually depends on having unobstructed buildable land, correlation of high powered winds with electrical needs, displacement of high priced purchased electricity, and a favorable price for excess electricity sold to the utility or others.

Probably the most relevant experience to installing and operating WECS by profit motivated firms is industry's major participation in cogeneration over the years. In fact, cogeneration and small power production (such as generating electricity by WECS) were grouped in the Public Utility Regulatory Policies Act of 1978 (PURPA) which gave the Federal Energy Regulatory Commission (FERC) the incentive to work with state public utility commissions and utilities to assure cooperation between on-site electricity generators and central station power producers.

Industry has a long history of both cogeneration (producing thermal and electrical power from common equipment) and generating electricity for its own use as well as sale to others. In 1980, industry generated 6000 MW of electrical power (5 percent of all electricity produced) by cogeneration, and DOE was predicting that capability would double by 1990. Those numbers are impressive, but back in 1900 industry produced more than 50 percent of the nation's electricity. By 1950 the portion had fallen off to 15 percent before dropping to the 1980 level of 5 percent.

That drop reflected a shift to reliance on utility-generated electric power, which looked more and more attractive as central station economies of scale and the development of modern power systems provided electricity at lower costs and higher reliability.

However, the Electric Power Research Institute (EPRI) recently suggested that with rapidly escalating oil and gas prices, the economics are now changing again. Because the cogenerator can usually produce electricity with less fuel per kWh generated (after credit for usable thermal energy) than the utility, oil and gas price increases have hit the utilities proportionately harder. The increased financial and environmental constraints on utilities, plus the uncertainty of long lead times for building new coal and nuclear generating stations, help explain why attention is once again on industry as a source of electric power. EPRI does indicate they expect a leveling off at an upper limit of 15 percent of total electricity by industry by the turn of the century.

On-site generation of electricity by WECS can be viewed in the same way as cogeneration. If the electricity can be used in on-site operations, purchased electricity can be reduced by utilizing the otherwise wasted windpower as a fuel. If the wind is available when not needed in operations, the resultant electricity can be sold to the utility. Feasibility depends on the relative costs and values which, in turn, depend on correlation and intensity of winds with peak loads, overall demand, and energy use patterns. These issues are discussed in more detail in the final chapter on project development.

With favorable implementation of PURPA, and the Economic Recovery Tax Act of 1981, there has been renewed interest in intermediate and large size WECS by industry. Other profit-oriented businesses and companies involved with major processes such as mining, oil and gas extraction and distribution, and water and sewage pumping and treating, have started to investigate opportunities to save or make money with wind power.

Generating power from the wind under PURPA looks like a plum for business and industry. Yet many managers express reservations about getting into the power business, even though PURPA's exemptions have now freed them from the long-held fear this would subject them to regulation as utilities. Producing electricity is, after all, not the business they know best. It could require a substantial investment in new plant and equipment, and it might require personnel with special skills and know-how that their own staffs may lack.

These concerns are realistic and good reasons why only projects with the best possible chance for technical and economic success should be undertaken. It is not an appropriate adventure for risk-adverse or timid companies, or those without the management or technical competence needed to pioneer new opportunities.

However, if you have land which is exposed to good winds (at least 14 mph measured at 30 ft) and if the cost of electricity to your business is at least $.06/kWh, you have the start of an attractive WECS installation. If you don't want to allocate the people resources to develop a WECS project yourself, consider joint venturing with your electrical utility (they can own up to 50 percent of the project) or making a deal with a professional windfarmer or other investors to allow use of your windy land in return for either a share of the project, payment for use of the land, or electricity generated by the WECS.

Electricity costs are expected to rise fastest where the utilities are most dependent on oil, natural gas, and fully developed water resouces as fuels. In parts of the Northeast, California, Hawaii, the Great Plains and the Pacific Northwest, those fast-rising costs correlate well with wind availability.

Approximately half or all non-urban industrial plants and processing facilities are believed to be located in areas with good winds. Energy needs are expected to grow at the same time energy costs are expected to rise. DOE has projected industrial use of energy will grow from 29 percent of all energy utilization in 1979 to 45 percent in 2000 and 50 percent in 2020. If the nation's economy and competitive position is to improve, industry must use more energy to provide more products. If wind, coal, nuclear or some other alternative fuel can't help keep the cost of energy at reasonable prices, the cost of products will have to rise and put U.S. industry at a competitive disadvantage in the world.

Not surprisingly, smaller industrial firms are located in the windy non-urban areas and often pay a premium for energy. Larger industrial plants have evolved in or near urban centers because many people are needed to operate those plants and processes and, historically, if people have had a choice they have chosen not to build communities in known high wind areas.

Commercial businesses have followed the same pattern as industry. Major businesses have headquartered and grown up in America's urban centers. Usually, those located in windy rural areas are smaller in size and less energy intense. However, studies for DOE indicate that there are over 100 thousand shopping centers, wholesale establishments, and warehouses located at sites with adequate winds to justify consideration of WECS. Although a small proportion of the nation's nine million commercial establishments, those businesses have the opportunity to save energy and money by effectively utilizing the wind as a fuel for electricity.

Although wind and buildable land availability are the two most important considerations in evaluating the feasibility of WECS, consideration is being given to installing WECS hardware on the roofs of some buildings to raise rotors higher into the wind and above obstructions while reducing the land area needed. However, careful engineering of the total system and close attention to safety is essential if large high-speed rotating equipment is to be installed near people. Early installations should be well protected from access by unauthorized people as well as sited in a location where failure of any parts of the machines will not result in danger to employees or the public.

In addition to the potential for economic and energy availability benefits WECS may provide, there is an excellent opportunity for favorable public relations, identification of product and business opportunities, and reducing dependence on imports of high priced fuels by installing WECS at your plant or business site.

A more detailed discussion of WECS project planning and economic analysis is included in the final chapter. However, a quick analysis checklist for business and industry is shown in Table 4-2.

GOVERNMENTS AND THE INSTITUTIONS

Public and not-for-profit agencies can look at WECS installations in a different light. Tax incentives are not applicable to direct purchases or installations if the potential owner is not a profit-oriented taxpayer. However, direct costs generally can be less because of lower cost of

Table 4-2. Quick Analysis Check List: WECS for Business and Industry.

Site Evaluation
    Is buildable land available?
    Do winds average over 14 mph?
    Is the site accessible to powerlines?

Electrical Evaluation
    What are electrical demands and costs?
        Peak kW power? Demand charge?
    Annual kWh energy usage?
        Cost and rate structure? Seasonal use?
        Time of day? Days of week?

Project Evaluation
    Do WECS fit with available resources?
    Is a consultant or partner needed?
    Will the utility cooperate?
        Interconnection? Buy-back rate? Invest?
    Are state or local incentives available?
        Tax credits? Loans? Other?
    Will WECS help my company?
        Economics? Community relations? Business opportunity?

money utilized or borrowed. In addition, the generation of profits along with energy is not needed. Much more emphasis can be placed on the values of community leadership; reduction of imported fuels and foreign reliance; utilizing the wind as a renewable, nonpolluting, generally benign, locally available fuel; and the educational and pioneering aspects of being involved at the beginning of something good for the country and the general population.

All that is in addition to a chance to save money over the years if WECS are installed at sites with good winds.

Major government (federal, state, and more local) facilities where WECS are worthy of consideration include:

- Military bases;
- Shipyards, supply depots, and maintenance areas;
- Prisons, health care and rehabilitation centers;
- Water and sewage treatment and pumping stations;
- Parks, forests, and other government lands;
- Highway and transportation facilities;
- Campuses for research, education, and office complexes;
- Community services (such as street lighting).

Many of these facilities have their own power-generating capabilities, usually diesel-driven electric generators or oil- and gas-fired steam generators. They usually are interconnected with the local utility.

Probably the most favorable conditions for WECS installations occur at military bases. Many bases generate at least some of their own power, often using oil or natural gas-fueled boilers or diesel generators. Most domestic bases are tied directly to the local electrical utility as well. The military also has technical and operations personnel competent to design, procure, install, operate, and maintain WECS. With the Washington mandate to dramatically reduce use of fossil fuels, installation of WECS at bases with good wind resources is a natural evolution.

As part of a "mission analysis" for the Energy Research and Development Administration (ERDA, later transformed into DOE) in 1977, General Electric found that four military facilities (three Air Force and one Navy) were located in areas with excellent wind resources (Class 7 power) and another 29 (13 Air Force, 10 Navy, and 6 Army) were located in areas with better than average winds (Wind Power Classes 4, 5, and 6), without counting the known windy military sites in Alaska, Hawaii, Puerto Rico, and other offshore locations.

Since that time, the Department of Defense (DOD) assigned different branches the lead role in various alternative energy technologies. The Air Force has primary responsibility for researching and reporting to the other services on both large and small WECS. Isolated Army bases also investigated large WECS and all branches investigated small systems. The Marines teamed with the Navy for their input. The Coast Guard concentrated mainly on small remote applications.

The initial technical responsibility for military WECS was assumed by the Air Force Academy at Colorado Springs. In addition to building and analyzing a very small (1.2 kW) VAWT, they developed a methodology to analyze potential wind sites, which included means to economically compare WECS power with other options.

Individual Air Force bases in known windy areas also proceeded with feasibility studies and one (Travis Air Force Base in Fairfield, California) proposed installing 4 MW of WECS capacity to supply as much as 40 percent of its power needs.

The U.S. Army activities were more decentralized and relied on the technology efforts of DOE and the Air Force. Grass roots activities at Fort Sill (Oklahoma) and Fort Ord (California) were productive. However, lack of funding kept the Army, as well as other military branches, from fulfilling their goal of reducing dependence on fossil fuels.

The Wind Energy Act of 1980 included specific provision for military and other federal government purchases of WECS at taxpayer-owned sites with 12 mph wind averages or better. When that law was not funded, many of the government's efforts with WECS were left in limbo. However, with the knowledge gained and the site investigations which were made, the opportunity exists to install WECS on an economically attractive basis without special subsidies or incentives if government agencies, utilities, and the private sector WECS providers can cooperate.

In addition to the military bases, any federal facility with buildable land and good wind resources should be proceeding with installation of WECS to fulfill its congressionally mandated leadership role in reducing use of depletable and expensive fossil fuels. Although the Wind Energy Act set 12 mph as the minimum winds needed for WECS, that minimum should be raised to 14 mph for economic viability unless subsidies are provided.

Early analysis indicated that as many as 175,000 out of 405,000 government buildings are located in areas with at least 12 mph wind speed averages. If the economically attractive base is raised to 14 mph, the building sites will probably be less than 50,000. However, even that number is more than can be served by available WECS in the '80s, and the sites with the best combination of wind resources, energy use patterns and cost, and compatibility with local utility conditions can be selected to allow government to lead in a responsible way.

STATE LEVEL LEADERSHIP

At the state level, many individual states proceeded with WECS programs without waiting for federal government help. Alaska, California, Hawaii, New York, Oregon, and Wisconsin took early proactive (beyond offering tax incentives) leadership positions, usually in cooperation with universities and utilities, to help remove obstacles and encourage responsible WECS installations. Much of the WECS actively in those states today is directly attributable to aggressive state policies of the '70s.

Other states wanting to get something going to help with fuel and electricity shortages and high costs are encouraged to combine the best of the California, Hawaii, and Oregon programs and to execute them with close cooperation between state energy offices, utility commissions, and individual utilities and their state associations. If any help comes forth from the federal level, accept it, but don't depend upon or wait for it.

The California Wind Energy Program

This program headed by the California Energy Commission (CEC, chaired by former astronaut Russell L. Schweickart) was initiated in 1977 by legislative action which established the goal that 10 percent of the state's electricity (approximately 7700 MW) be provided by the wind by the end of the century. Specific Energy Commission assignments were:

- Assessment of wind resources throughout California;
- Operation of a public wind information center;
- Testing of electricity generating WECS;
- Research leading to the development of large-scale prototype wind turbines suitable for California.

CEC has aggressively pursued those responsibilities in cooperation with many other state agencies and investor-owned and municipally owned utilities. The involved state agencies include:

- Office of Appropriate Technology;
- Statewide Energy Consortium, the California State University and Colleges;
- Public Utilities Commission;
- Department of Water Resources;
- Department of Food and Agriculture;
- California Conservation Corps;
- Department of Parks and Recreation;
- Department of Business and Transportation;
- California Department of Transportation.

Early utility leadership in WECS development programs was provided by:

- Southern California Edison Company (SCE);
- Pacific Gas and Electric Company (PG&E);
- Los Angeles Department of Water and Power;
- City of Santa Clara Municipal Utility.

In addition to valuable impetus, wind resource assessment and mapping, and WECS data dissemination, the California program evolved major WECS demonstration programs in the San Gorgonio Pass, near

Palm Springs (SCE's WECS research site with Bendix/Schlachle 3 MW HAWT, Alcoa 500 kW VAWT and DAF-Indal 50 kW VAWT proto-types); in Solano County, forty miles northeast of San Francisco (PG&E's research project with a DOE/NASA MOD-2, 2.5 MW HAWT, built for PG&E by Boeing Engineering & Construction Co.); and at the San Luis Reservoir, near Los Banos (California Department of Water Resources' pilot WECS test project in cooperation with the Office of Appropriate Technology, Alcoa, and PG&E, with a DAF-Indal 50 kW VAWT installed by K.G. Walters Construction Company).

The Hawaii WECS Program

This program was inspired by the excellent winds and almost com-plete (90 percent) dependence on imported oil for electricity to serve the state's dynamic growth. The Hawaii Natural Energy Institute (HNEI) at the University of Hawaii at Manoa (UHM) has been the lead agency in organizing state resources in support of a WECS con-tribution to the Islands. D. Richard Neill has been a driving force as Wind Program Coordinator, and has been unswerving in his belief that "Hawaii has some of the most favorable wind regimes in the world."

With northeasterly trade winds blowing across the state approxi-mately 70 percent of the time (at mean wind speeds in the 14–24 mph range), Neill early classified Hawaii's WECS potential as enor-mous. A 1981 HNEI "Report on Hawaii's Wind Energy Potential" quantified enormous into a low to high estimate of WECS capable of generating 27 billion to 131 billion kWh per year (based on windy land coverage at typical wind farm rotor spacings, but without regard for availability or buildability of that windswept land) which is 4–21 times the electricity sold on the five islands in 1979.

For perspective (but certainly not as a projection), the HNEI re-port related that amount of land and energy generation to the WECS hardware equivalent of 289–723 Hamilton Standard WTS-4 (4 MW) HAWTs, 201–500 DOE/NASA MOD-2 (2.5 MW) HAWTs, or 2893–7232 Alcoa 123 ft × 82 ft (500 kW) VAWTs.

The total Hawaii program has been designated WEAN (Wind Energy Application Network) and is in cooperation with many public and private groups. It is intended to help wean Hawaii from its petroleum dependence by focusing on:

- Wind energy resource assessment;
- WECS applications research;
- WECS reliability verification;
- Nontechnical development issues.

The most important element of the WEAN program is a comprehensive *wind energy resource assessment* to insure intelligent hardware purchases and siting decisions. A premium is placed on identifying strong wind sites. Because wind measurements made at one location are not necessarily applicable to sites more than a few metres away, a comprehensive method for mapping the state's wind resources was followed which combined collecting long-term data from fixed reference stations and short-term data from mobile equipment. Several long-term monitoring stations were established on each island and include eight 150 ft meteorological towers instrumented at 30 ft, 90 ft, and 150 ft elevations by HNEI, DOE/PNL, and Maui County to record windspeed and direction. Additional stations are maintained by the U.S. Weather Service, UHM Department of Meteorology, HNEI, the U.S. military and other organizations. Four mobile vans, equipped with anemometers and recording instruments and operated by the UHM Department of Meteorology, are used as a mobile station sampling system. Thirteen TALA kites, eight of which have been modified to feed data directly into the mobile vans' microloggers, along with five which are hand held, also give short-term information on gusting, wind shear, and turbulence, as well as windspeed at higher elevations. In addition, HNEI maintained 20 instantaneous windspeed indicators and 20 wind data accumulators (WDA) for loan to those interested in wind energy conversion applications. Long-term mean annual windspeeds can be estimated for the site by comparing WDA results and correlating them with long-term data from the appropriate reference station. Annual energy output for a specific WECS at the site can be approximated from the WDA wind data and performance curves provided by WECS manufacturers, or developed from actual measurements from installed WECS. HNEI established a Wind Data Bank to store and make available data to public and private decision makers.

Although HNEI acknowledged that major use of WECS will be in generating utility interconnected electricity, *WECS applications research* was involved with direct waterpumping and irrigation applications, innovative wind-waterpumping devices, wind-nitrate fertilizer production, wind-hydrogen, wind-electric vehicle powering, and wind-battery storage. Other applications for WECS which were investigated include pump-storage hydropower systems, ventilation of tunnels, thermal energy, powering waste treatment plants, ice making and space cooling, salt water desalination, inorganic chemical production, and industrial processing of manganese nodules.

Because of Hawaii's leadership role in utilization of wind energy, its excellent wind resources made possible a *WECS reliability verification* program based on securing accelerated operating experience for WECS development prototypes and obtaining long-term reliability data on a variety of existing WECS. Putting WECS into service in favorable regimes helps resolve questions of economic viability, corrosion protection, optimization of output, and issues such as grid stability, safety, grid interface, and energy storage. Wind application data collected from installed WECS is fed into the HNEI computer for future evaluation of long-term operational results. WECS are instrumented to record windspeed, rotor rpm, amps, volts, VARS, kilowatts, and kilowatt hours.

The WEAN *nontechnical development issues* program is intended to ensure responsible utilization of Hawaii's windpower potential. It includes:

- Wind energy information dissemination:
  - available WECS and their performance,
  - wind data, general and site specific (wind energy maps and Hawaii Wind Data Bank),
  - application data,
  - wind measuring equipment loan program;
- Decision making assistance:
  - economic analysis tools,
  - wind energy potential analysis,
  - financing alternatives guidance,
  - government incentive programs identification;

- Identification of potential constraints:
  - social,
  - environmental,
  - aesthetic,
  - legal,
  - government requirements (zoning, approvals, etc.),
  - utility concerns;
- Development of needed technical infrastructure (training):
  - design,
  - installation,
  - operating and maintaining.

The favorable winds, aggressive HNEI actions, cooperative island utilities, a state tax credit of 10 percent (without upper limits), and a long standing state equivalent of PURPA (originally to cover sugar industry cogeneration with bagasse) has allowed Hawaii to become a leader in WECS installations, summarized as follows:

- DOE/NASA MOD-0A (200 kW HAWT) at Kahuku (Oahu) began operating in June, 1980 in cooperation with Hawaiian Electric Company (HECO) and is known affectionately as "Makani Huila" (windwheel);
- Mehrkam 20/40 kW HAWTs on Molokai, by Molokai Energy Company, and on Hawaii, by Kahau Ranch;
- Millville 10 kW HAWT on Molokai by Hawaiian Homes Commission;
- ESI-54 (50 kW HAWT) prototype on Molokai by Molokai Energy Company;
- Jacobs 10 kW HAWTs on Hawaii.

In addition, wind farms have been announced by Windfarms, Ltd. (on Oahu, Maui, and Hawaii), Molokai Energy Company (on Molokai) and Kahua Ranch (on Hawaii) in cooperation with the local investor-owned utilities.

The Oregon Experience

Oregon's adventures in winds are unique, in that the Northwest has long enjoyed electricity prices among the lowest in the United States. Low cost hydroelectric power distributed by Bonneville Power Administration (BPA) was being sold retail in many communities in Oregon for less than 2¢/kWh in 1981 while the national average was approaching 6¢. In addition to being accustomed to low cost energy, Oregon's people are among the most conscious and protective of their environment and many have worked to "keep Oregon green." When it became obvious that population and energy growth in the Northwest was outstripping the fully developed hydroelectric capacity, environmentally benign alternatives (such as wind and solar) became attractive. When BPA set the early '80s for both price increases and limits on supply growth, the local power distributors (such as municipal utilities, rural electric cooperatives, and people's utility districts) had to initiate action, even though they didn't expect to find competitive cost options. Even BPA expanded its horizons by encouraging dispersed power production interconnected with its system.

The Oregon wind program got underway in the mid-70s with one of the nation's first anemometer loan programs and an effective tax credit incentive program. The state legislature later authorized creation of the Oregon Alternative Energy Development Commission, which was established with a wind task force in the late '70s. That 12-person task force was headed by Nicholas Butler of Bonneville Power Administration (BPA, headquartered in Portland) and ramrodded by Donald Bain, wind specialist for the Oregon Department of Energy (ODOE). The task force members represented private and public utilities, WECS distributors and dealers, academia, and government agencies.

The wind energy task force concluded that wind was, in fact, a leading practical alternative for Oregon and estimated that, by the end of the century, larger WECS could supply power the equivalent of over 1300 MW of thermal power plants, and smaller dispersed WECS could contribute at least another 25 MW. The task force recommended encouraging and publicizing installation of promising WECS hardware in windy portions of the state, and made recommendations to the Alternative Energy Commission for specific actions, many of which have been implemented by Oregon's legislative and executive branches.

The Oregon DOE, led by Don Bain, proceeded with an effective statewide WECS program to:

- Administer tax credits;
- Provide technical assistance to other Oregon entities:
  - utilities,
  - local governments,
  - business,
  - the public;
- Provide public information:
  - publications,
  - films,
  - referrals,
  - group presentations around Oregon;
- Gather information on developing markets and applications and WECS performance;
- Assist with consumer protection issues and conflict resolution;
- Administer the anenometer loan program;
- Review regional and specific utility planning relative to wind energy utilization;
- Manage special WECS programs relative to:
  - wind resource assessment,
  - wind rights,
  - land use planning,
  - permitting processes.

In addition to major BPA wind programs (such as hosting the DOE/NASA experimental MOD-2 wind farm—three large HAWTs totaling 7.5 MW capacity—and several dispersed small WECS on the Washington side of the windy Columbia River Gorge) most of the investor-owned and publicly-owned utilities in Oregon got involved in important WECS programs.

Portland General Electric (PGE) participated on the wind task force, funded wind measurement and analysis projects, and cooperated with their customers interested in installing small dispersed WECS intertied with their distribution lines.

Pacific Power and Light (PP&L, headquartered in Portland) also undertook wind resource assessment activities, both on their own and in cooperation with Oregon State University. As soon as they found good winds, PP&L proceeded to install and test a prototype 300 kW capacity HAWT (WTG Energy Systems MP3-200) at Whiskey Run (on the coast near Coos Bay) as a major step in their research program.

Another ambitious WECS research project was initiated in 1978 by two publicly owned utilities, Eugene Water and Electric Board (EWEB) and Central Lincoln Peoples Utility District (CLPUD). Oregon's public utilities had earlier worked with the wind research staff at Oregon State University to identify good wind sites and were eager to find alternatives to historic BPA supply in the '80s. Keith Parks, general manager of EWEB, and Harold Sudduth, then general manager of Central Lincoln, decided it was time for hands-on experience and proceeded to involve the American Public Power Association (APPA) Research Committee, Alcoa, and a majority of the Oregon municipal, PUD, and rural electric cooperative utilities to fund and build an intermediate-size WECS for demonstration and testing on the windy coast. Costs were shared equitably by all the participants, which included:

- Aluminum Company of America (Alcoa), system hardware;
- American Public Power Association, research;
- Bandon Utilities;
- Blachly-Lane Cooperative;
- Canby Utility Board;
- Central Lincoln PUD, host utility;
- City of Ashland;
- City of Forest Grove;
- City of McMinnville;
- Clatskanie PUD;
- Coos-Curry Electric Cooperative;
- Drain Light & Power;
- Douglas Electric Cooperative;
- Eugene Water and Electric Board, project manager;

- Lane Electric Cooperative;
- Midstate Electric Cooperative;
- Milton-Freewater Light & Power;
- Northern Wasco County PUD;
- Salem Electric;
- Springfield Utility Board;
- Tillamook PUD;
- Umatilla Electric Cooperative;
- West Oregon Electric Cooperative.

That cooperative research venture led to the installation in 1980 of the world's largest vertical axis wind turbine (123 ft high by 83 ft diameter rotor, 500 kW capacity) at a windy site above Agate Beach (near Newport) overlooking the Pacific Ocean. After testing and later retrofitting, that VAWT served as a working prototype of Alcoa's ALVAWT Model #1238229–500 kW.

Local Government: Livingston, Montana

A pioneering local government adventure in windpower is underway by the city of Livingston, Montana. That city found it was "the seventh windiest reporting location in the continental U.S.A." and decided to develop and market its enormous wind energy potential.

Livingston is located at the western end of the Livingston Bench, where classic windpower topography and climate converge to provide average windpower density of over 500 watts per vertical square meter (Power Class 6). The area is 56 miles north of Yellowstone National Park and prevailing winds are funneled between mountains northeasterly along the Yellowstone River wind corridor.

In cooperation with Montana state agencies, the U.S. Environmental Protection Agency, DOE, and Montana Power Company, Livingston proceeded with:

- Partial powering of the municipal sewage treatment plant with five 25 kW HAWTs;
- Wind monitoring of a candidate site for a possible large DOE prototype installation;

- Instrumentation and monitoring of the correlation between winds and electrical output of one of the 25 kW HAWT prototypes;
- Installation of two 10 kW residential HAWT systems;
- Installation, instrumentation, and testing of a very small HAWT as part of the DOE field evaluation program;
- Consideration of a 90 MW wind farm.

With that impressive start, Livingston's Community Development Office (with Ed Stern as Director) proceeded to package available financial and developmental incentives, prepare an impressive brochure to introduce the opportunity, and aggressively solicit participation by WECS project developers and producers.

Productive efforts, such as those described in California, Hawaii, Oregon, and Livingston, have provided the basis for effective cooperation between all levels of government, WECS providers, and WECS users to harness the previously wasted power of the wind and turn it into useful electricity. That type of cooperation and the resultant utilization of an abundant renewable energy resource will ultimately be beneficial to all Americans.

5
Project Planning and
Development

Developing a WECS project can be complex and technically demanding. If there is no local experience to build on, it probably will be even more challenging. It is not an appropriate activity for risk-averse companies or managers. At this early stage of WECS industry evolution, a little pioneering spirit and entreprenuerial flair will be helpful. The ability to initiate, effectively plan and execute a multi-faceted project will require blending of many management and professional skills.

This chapter deals with the issues of deciding whether a WECS project should be undertaken and, if so, how to do it. The emphasis is on determining economic feasibility and, under favorable conditions for development, minimizing risks and maximizing financial rewards.

The most important factors for success in a WECS venture are site characteristics, public considerations, future value of energy, WECS performance capability and reliability, and effective execution of a competent action plan. Those factors are covered in the following general categories:

- Wind resources and siting;
- Local wind analysis;
- Consideration of the public;
- Project planning and analysis;
- WECS project development.

WIND RESOURCES AND SITING

Selecting a site, or determining if an available site is suitable, for WECS involves most of the issues relative to any kind of construction project, except that availability of adequate winds is added as number one on the list. Without good winds, there are no other issues.

There is little agreement on how much wind is needed to fuel a successful WECS project and, in fact, the minimum will vary with competitive electricity and fuel costs and will depend on specific WECS performance and cost. However, since wind power is proportional to the cube of windspeed (all other factors in the conversion formula are linear) the wind resource available to the WECS rotor is by far the most important consideration.

The ill fated (never funded and therefore with no redeeming social importance) U.S. Wind Energy Systems Act of 1980 was intended to provide subsidies, direct government purchases, and other incentives to stimulate WECS projects so that wind could make a significant contribution to the nation's energy needs. Even with its positive thrust, and the normal congressional desire to include and please as broad a constituency as possible, the act stipulated that sites with winds averaging at least 12 mph be utilized. Considering the momentum and political thrust of the times, the 12 mph minimum has to be considered optimistic. In the real world of the '80s, any known combination of anticipated WECS hardware and competitive domestic energy costs will require at least 14–15 mph average winds to make a WECS installation economically attractive.

Therefore, the first (and perhaps only) question is, does your site have adequate winds, or can you get control of one that does? To help answer this question, the Federal Wind Energy Program has had an active "wind characteristics program element" since 1976. Pacific Northwest Laboratory (PNL, run for DOE by Battelle Memorial Institute in Richland, Washington) has had responsibility for locating, understanding, and mapping the wind resources of the United States and providing useful interpretations of that data to anyone who can put it to use. Their work culminated in 1981 with the publishing and distribution of a national wind resource atlas and 12 regional wind atlases. Those publications depict in graphic, tabular, and narrative forms wind resources at the regional and state levels.

For their purposes in mapping wind resources, PNL chose to utilize "windpower density" instead of windspeeds so that a single number could represent the combined effect of windspeed distribution, air density, and the effect of cubing the windspeed. The PNL data present windpower at two typical heights above the ground (33 ft and 164 ft, or 10 m and 50 m) based on extrapolation of measured windspeeds at any available height to those two reference heights by the commonly utilized 1/7 power law. Table 5-1 defines seven standard windpower classes utilized in all of the DOE/PNL wind resource assessment work.

Wind data used in the assessments were obtained from sources such as the National Climatic Center (NCC), the U.S. Forest Service, universities, utilities, and other government and private organizations. The breadth of windspeeds within power classes reflects the lack of both quantity and quality of historic wind data and allows presentation without overly implying accuracy. The DOE/PNL atlases provide a good starting point for determining geographic areas for consideration of WECS installations.

If you accept the conclusion that at least 14–15 mph average winds will be needed, look for sites with Power Classes 5, 6, or 7. Yes, the 14–15 mph must be found at the lower height. The winds are expected to be better higher. Figure 5-1 addresses the relevance of height to expected speeds and power densities. Two additional important assumptions, Rayleigh windspeed distribution and air density, are discussed in Figures 5-2 and 5-3.

Table 5-1. Windpower Classes as Utilized by the Department of Energy and Pacific Northwest Laboratory.

WIND POWER CLASS	10 M (33 FT)		50 M (164 FT)	
	WIND POWER DENSITY, WATTS/SQ. M	MEAN WINDSPEED M/S (MPH)	WIND POWER DENSITY, WATTS/SQ. M	MEAN WINDSPEED M/S (MPH)
1	0	0	0	0
2	100	4.4 (9.8)	200	5.6 (12.5)
3	150	5.1 (11.5)	300	6.4 (14.3)
4	200	5.6 (12.5)	400	7.0 (15.7)
5	250	6.0 (13.4)	500	7.5 (16.8)
6	300	6.4 (14.3)	600	8.0 (17.9)
7	400	7.0 (15.7)	800	8.8 (19.7)
	1000	9.4 (21.1)	2000	11.9 (26.6)

Based on Rayleigh windspeed distribution and standard sea-level conditions. To maintain the same power density, speed increases approximately 5 percent per 5000 ft (3 percent per 1000 m) of elevation.

HEIGHT (FEET) ABOVE GRADE	WINDSPEED HEIGHT ADJUSTMENTS; MULTIPLY 33 FT WIND-SPEED BY: SURFACE ROUGHNESS (EXPONENT)		
	SMOOTH SAND (.1)	COMMON (.14)	ROUGH FOREST (.3)
40	1.03	1.04	1.09
50	1.05	1.07	1.16
60	1.07	1.10	1.23
70	1.09	1.13	1.29
80	1.10	1.15	1.34
90	1.12	1.17	1.39
100	1.13	1.18	1.44
120	1.15	1.21	1.51
150	1.18	1.25	1.62

Figure 5-1. Correction of wind data for variations in rotor height and surface conditions.

Considerable valuable information is available in the DOE/PNL atlases and in related publications. A quick screening of those publications identified 27 states in the continental U.S. with Class 5 or better wind sites. (See Table 5-2.) Of those, 20 have some Class 6 or better, including four with Class 7 windpower. Although the percentage of total land is small, in just those 27 states there are more than 31,000 square miles of land believed to have access to Class 5 or better winds, with 4200 square miles of that having Class 6 plus and almost 400 square miles exposed to Class 7 winds. Although that is more than enough land to site all of the WECS hardware which could possibly be built in the next few decades, much of it is unavailable, covered by lakes or rivers, or in remote or rugged areas too difficult to build on. Still, there is plenty left for billions of dollars of WECS developments.

Outside the lower 48, wind resources are even more dramatic. (See Table 5-3.) Those windy areas add another 35,000 square miles to the Class 5 or better total, including over 24,000 square miles better than Class 6, with 10,630 square miles in Class 7.

To put those real estate numbers into WECS power perspective, at a modest 100 MW per square mile there is a land-limited capacity of 320,000 MW of installed WECS generating capacity, without including Alaska. Class 7 sites alone can accept over 40,000 MW. Alaska is excluded from these totals only because it has far more windy acreage than all the rest of the states and territories together. It deserves separate attention.

WINDSPEED		HOURS PER YEAR IN AVERAGE OF:		
M/S	MPH	5.4 M/S (12 MPH)	6.7 M/S (15 MPH)	8.0 M/S (18 MPH)
4.5	10	554	431	333
4.9	11	543	441	348
5.4	12	523	443	359
5.8	13	494	441	367
6.3	14	459	432	370
6.7	15	420	418	369
7.2	16	378	400	365
7.6	17	336	379	358
8.1	18	294	355	348
8.5	19	253	330	336
8.9	20	216	303	322
9.4	21	181	275	306
9.8	22	150	248	288
10.3	23	123	222	271
10.7	24	99	197	252
11.2	25	79	173	233
11.6	26	62	150	214
12.1	27	48	130	196
12.5	28	37	111	178
12.9	29	28	94	160
13.4	30	21	79	144
13.9	31	16	66	128
14.3	32	11	55	114
14.8	33	8	45	100
15.2	34	6	37	88
15.6	35	4	30	76
16.1	36	3	24	66
16.5	37	2	19	57
17.0	38	1	15	49
17.4	39	.9	12	41
17.9	40	.6	9	35
18.3	41	.4	7	30
18.8	42	.3	5	25
19.2	43	.2	4	21
19.7	44	.1	3	17
20.1	45	–	2	14
20.6	46	–	2	12
21.0	47	–	1	9
21.4	48	–	.9	8
21.9	49	–	.7	6
22.3	50	–	.5	5

Figure 5-2. Rayleigh distribution of windspeeds for 12, 15, and 18 mph wind regimes.

ALTITUDE		TEMPERATURE		
(FT)	MULTIPLY BY	°C	°F	MULTIPLY BY
Sea level	1.00 (base)	–17.8	0	1.13
1000	0.97	– 6.7	20	1.08
2000	0.93	4.4	40	1.04
3000	0.91	15.6	60	1.00 (base)
4000	0.88	26.7	80	.96
5000	0.83	37.8	100	.93
7500	0.76			
10000	0.69			

Figure 5-3. Correction of wind data for variations in air density caused by altitude and temperature.

East of the Mississippi the powerful winds are found mostly in the mountains, along the Atlantic coast, and in parts of the Caribbean. Only 11 states (Connecticut, Maine, Massachusetts, New Hampshire, New York, North Carolina, Rhode Island, Tennessee, Vermont, Virginia and West Virginia) have significant land with Class 5 winds or better, and none has significant land area with Class 7 power.

The best wind resources in New England are in the White Mountains of New Hampshire, the Green Mountains of Vermont, Maine's Longfellow Mountains, and the coastline and islands of Maine and Massachusetts. The Mt. Washington Observatory in New Hampshire claims a record with a windspeed recording of 231 mph.

New York State has patches of windpower in the Adirondacks, the Catskills, on North Mountain, on the eastern tip of Long Island, and in a small area above Elmira in the southern Finger Lakes region.

The Southeast is calm with the exception of the mountainous area at the North Carolina–Tennessee border and up along the Appalachian chain, through small parts of Virginia and West Virginia. The Great Smoky, Blue Ridge, and Allegheny Mountains appear to hold the promise of the best winds in the Southeast.

Mid-America, from Mexico to Canada and including the Great Plains and Rocky Mountains, is notorious for its winds. Eleven states (Arkansas, Colorado, Kansas, Montana, New Mexico, North Dakota, Oklahoma, South Dakota, Texas, Utah, and Wyoming) have at least a trace of Class 5 windpower and two (Texas and Wyoming) have

Table 5-2. Wind Resources in Continental United States.

STATE	5 & BETTER		6 & BETTER		7	
	%	AREA (SQ. MI.)	%	AREA (SQ. MI.)	%	AREA (SQ. MI.)
Arkansas	0.1	30	—	—	—	—
California	0.5	900	0.2	300	0.1	60
Colorado	1.2	1,300	0.1	40	—	—
Connecticut	0.1	4	—	—	—	—
Idaho	1.8	1,500	0.4	300	—	—
Kansas	2.0	1,600	—	—	—	—
Maine	1.6	500	0.4	140	—	—
Massachusetts	0.9	120	<.1	2	—	—
Montana	3.9	6,000	0.7	1,000	—	—
Nevada	1.5	1,500	0.7	800	0.1	108
New Hampshire	2.5	220	0.1	10	—	—
New Mexico	0.9	1,100	0.1	40	—	—
New York	0.3	150	<.1	10	—	—
North Carolina	0.3	150	<.1	20	—	—
North Dakota	0.5	350	—	—	—	—
Oklahoma	3.8	2,500	0.1	80	—	—
Oregon	0.2	180	<.1	10	—	—
Rhode Island	1.0	4	<.1	1	—	—
South Dakota	0.2	180	—	—	—	—
Tennessee	<.1	15	<.1	5	—	—
Texas	3.0	8,000	0.2	400	0.1	150
Utah	0.3	220	0.1	40	—	—
Vermont	2.5	220	0.1	10	—	—
Virginia	0.1	15	—	—	—	—
Washington	0.5	300	<.1	12	—	—
West Virginia	0.4	100	—	—	—	—
Wyoming	4.4	4,400	1.1	1,000	0.1	80

Table 5-3. Wind Resources in Alaska, Hawaii, and the U.S. Territories.

	WIND POWER CLASS:					
	5 & BETTER		6 & BETTER		7	
STATE POSSESSION	%	AREA (SQ. MI.)	%	AREA (SQ. MI.)	%	AREA (SQ. MI.)
Alaska	6.1	35,000	4.1	24,000	1.8	10,000
Guam	1.4	3	—	—	—	—
Hawaii	3.9	240	2.8	160	1.0	60
Marshalls	18.3	10	8.1	6	4.9	3
Midway/Wake	62.3	4	—	—	—	—
Puerto Rico	0.1	2	—	—	—	—

more than half of the lower 48's Class 7 land. However, with the exception of small areas in southwest Texas (Davis Mountains and Guadalupe Pass) and the Sacramento Mountains in southern New Mexico and Ouachita Mountains at the Arkansas–Oklahoma border, the bulk of the powerful winds are mainly in the plains, and in the Rocky Mountain states from Texas and New Mexico northwest into Canada.

The high plains area (where Texas, Oklahoma, New Mexico, Colorado and Kansas get together) is one of the nation's prime wind targets.

Wyoming's south central plains are infamous for their winds and one of the nation's major wind energy research projects is underway in the Medicine Bow area. In addition, practically all mountain areas in western Wyoming and Montana have high windpower potential.

The West, including Alaska and the U.S. interests in the Pacific, not only has an abundance of good, better, and best wind resource locations, but is also the part of the country most studied, prospected, and apparently committed to development of its wind resources. Ten states and possessions (Alaska, California, Guam, Hawaii, Idaho, the Marshall Islands, Midway and Wake Islands, Nevada, Oregon, and Washington) have good Class 5 winds, and five (Alaska, California, Hawaii, the Marshalls, and Nevada) have Class 7 power.

Most of Alaska has outstanding winds, with south central and southwest areas having high percentages of land exposed to Class 7 winds. The Alaska Islands, the coast of the Beaufort, Chukchi, and Behring Seas, the Alaskan peninsula, Aleutian Islands, Lower Cook Inlet, Gulf of Alaska coast, the mountains of the western Alaska range and the eastern end of the Brooks Range all appear to have excellent wind resources.

Parts of all the Hawaiian Islands are excellent, as are the Marshall Islands. Over 60 percent of Midway and Wake Islands are swept by Class 5 windpower.

On the mainland, the Pacific Northwest has good winds along the coast, on both sides of the Columbia Gorge, and in the Cascades. Major WECS R&D projects are underway in Oregon and Washington.

California and Nevada have large areas of Class 7 windpower in many different locations. California has studied its resources in more detail than any other state, and has many active wind energy projects underway. Better known locations include the coastal gap areas at the Carquinez Straits, Altamont Pass, Pacheco Pass, San Gorgonio Pass, and the Sierra Pelona.

All broad identification of locations with good probability of adequate wind resources should do is encourage or discourage further analysis, or guide you toward the better of alternative sites which may be available. Ultimately, site-specific wind analysis will be needed as part of a broader and more detailed feasibility study.

LOCAL WIND ANALYSIS

Because of the dominating importance of wind availability to consideration of any WECS, a good understanding of that wind resource is needed at the location where WECS are expected to be installed. Because the windspeed is cubed in formulae for predicting the ability of WECS to convert the kinetic windenergy to useful electrical energy, a small error in estimating the wind regime in which the WECS must perform can result in major variations in the amount of electricity generated by the WECS over its lifetime. Even though the wind is a free fuel, it is the most important economic consideration.

If several locations are available, the wind resource at each should be analyzed in enough detail to allow comparison between those sites. The amount of detail and accuracy, as well as requirements for professional skills, will vary with the stage of the investigation and the seriousness of intent to develop economically attractive WECS installations.

Because wind resources are so variable, and because the cost of site-specific analysis can be high, there is considerable interest in estimating wind energy potential without having to make on-site wind measurements. Several techniques can be used to screen candidate resource areas for sites with high potential or to estimate wind energy characteristics at a specific location. Useful indicators of wind potential are:

- Data prepared by others;
- Topographical evidence;
- Other visual site phenomena;
- Social and cultural awareness;
- Measuring and recording the wind.

Gathering wind data prepared by others is so obvious that it is often overlooked. In addition to public data generated specifically for WECS projects (available through normal DOE outlets and most state energy offices) other wind data can be found at state highway departments in areas where snow is expected; at public, private and military airports; at forest and park service firewatch stations; at Coast Guard stations; at pollution monitoring stations; at many radio and television stations; at many colleges and universities; and from wind energy dealers, consultants, prospectors, and WECS owners. If WECS are already in service in an area of interest, wind data from their performance history should be especially valuable.

Properly used, observing *topographical indicators* is cost-effective in shortcutting the siting process. In the past, interpretations of flow over terrain have been applied to siting wind machines without additional supporting measurements. Estimating topographic effects on windflow (such as the acceleration of wind over a ridge) is the oldest technique of wind energy site assessment, and is still valuable today.

In the resource evaluation stage of wind prospecting, topographical guidelines can determine locations for anemometers for confirming measurements throughout an area. These anemometers can easily be relocated if the incoming data indicate the initial anemometer siting was not optimum.

When screening candidate sites, topographical indicators can help distinguish a primary from a secondary site. If two candidate sites for WECS clusters have roughly equal windpower potential, indicators might suggest that one site experiences higher levels of turbulence than the other. That conclusion might result from the shapes of upwind topographic protrusions.

The following features are believed indicative of high mean windspeeds:

- Gaps, passes, and gorges in areas of frequent strong pressure gradients;
- Long valleys extending down from mountain ranges;
- High elevation plains and plateaus;
- Plains and valleys with persistent strong downslope winds associated with strong pressure gradients;
- Exposed ridges and mountain summits;
- Exposed coastal sites.

Features that signal low mean windspeeds are listed as:

- Valleys perpendicular to the prevailing winds;
- Sheltered basins;
- Short or narrow valleys and canyons;
- Areas of high surface roughness, such as forested hilly terrain.

Using *visual evidence of wind* ranges from simply watching the wind's actions over time or observing long-term effects of wind on the land or its vegetation. Because almost everyone tends to overestimate local winds (probably because miserably windy days are the most memorable) the Beaufort scale of windspeeds (See Table 5-4), prepared in 1806 for use at sea and later extended to use on land, can serve as a convenient reference point.

Although it is difficult to turn short-term observations into an understanding of long-term winds, recalling and relating the windspeeds associated with the Beaufort scale may help in determining whether an area should be investigated further.

Observing such longer term conditions as wind-formed sand dunes, wind-deformed vegetation, placement of snow fences, special windbreaks for pedestrian or vehicle protection, and special wind standards for design of buildings can provide further indication of direction and intensity of winds.

Experienced wind prospectors put considerable faith in studying tree deformation for determining prevailing wind direction, identifying areas where severe wind or ice loads may occur and, in fact, for estimating mean annual windspeeds. Although subject to uncertainty and requiring qualitative judgments, it is often possible to estimate average windspeeds within 20 percent just from close observation of trees. The Griggs–Putnam Index (developed in the 1940s by Palmer Putnam and botanist R.F. Griggs as part of their pioneering wind efforts in New England) is the best known attempt to develop a subjective scale. With comments added to relate to today's WECS, their eight classes, based on the degree of response of coniferous trees to the wind, are shown in Table 5-5.

Although all of the four most dramatic wind indicators (complete flagging through carpeting) offer encouragement that winds should be adequate to effectively fuel WECS, the two most severe also suggest special structural consideration when choosing or preparing a site.

Table 5-4. Modified Beaufort Scale of Windspeeds.

BEAUFORT NUMBER	APPROXIMATE WIND SPEED				DESCRIPTIVE TERM		VISUAL EVIDENCE
	MPH	M/S	KM/H	KNOTS	SEAMAN'S	WMO (1964)**	
0	<1	<0.3	<1	<1	calm	calm	Calm; smoke rises vertically.
1	1-3	0.3-1.5	1-5	1-3	light air	light air	Smoke drift indicates wind direction; vanes do not move.
2	4-7	1.6-3.3	6-11	4-6	light breeze	light breeze	Wind felt on face; leaves rustle; vanes begin to move.
3	8-12	3.4-5.4	12-19	7-10	gentle breeze	gentle breeze	Leaves, small twigs in constant motion; light flags extend; hair is disturbed, clothing flaps.
4*	13-18	5.5-7.9	20-28	11-16	mod. breeze	mod. breeze	Dust, leaves & loose paper raised up; small branches move, hair disarranged.
5*	19-24	8.0-10.7	29-38	17-21	fresh breeze	fresh breeze	Small trees in leaf begin to sway; force of wind felt on body.
6*	25-31	10-13.8	39-49	22-27	strong breeze	strong breeze	Larger branches of trees in motion; whistling heard in wires; some inconvenience in walking.
7*	32-38	13.9-17.1	50-61	28-33	mod. gale	near gale	Whole trees in motion; resistance felt in walking against wind.
8*	39-46	17.2-20.7	62-74	34-40	fresh gale	gale	Twigs & small branches broken off trees; walking generally impeded.
9	47-54	20.8-24.4	75-88	41-47	strong gale	strong gale	Some damage occurs; slate blown from roofs; danger of being blown over.
10	55-63	24.5-28.4	89-102	48-55	whole gale	storm	Trees broken or uprooted; considerable damage occurs.
11	64-72	28.5-32.6	103-117	56-63	storm	violent storm	Very rarely experienced on land; usually accompanied by widespread damage.
12-17	73-136	32.7-61.2	118-220	64-118	hurricane	hurricane	

*Within general operating range of most WECS. Conditions above Beaufort number 8 are considered dangerous and are more costly than helpful to WECS installations. Conditions represented by Beaufort numbers below 4 don't contain enough power to be of use.

**World Meteorological Organization. Since 1966, knots have replaced Beaufort numbers as weather map symbols.

Table 5-5. The Griggs-Putnam Index.

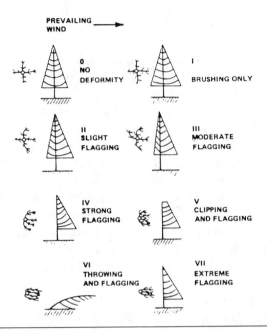

- Class O, No Effect. The wind has had no noticeable influence on the tree.
- Class I, Brushing. Small branches and needles appear bent away from the prevailing wind direction. The tree crown may appear slightly asymmetrical. Probably 6–10 mph wind averages.
- Class II, Light Flagging. Small branches and the ends of the larger branches are bent by the wind, giving the tree a noticeably asymmetric crown. Mean windspeeds in the 8–12 mph range.
- Class III, Moderate Flagging. Large branches are bent toward the leeward side of the tree, giving the tree a nearly one-sided crown. Almost good enough for WECS, 11–15 mph average winds.
- Class IV, Strong Flagging. All the branches are swept to the leeward and the trunk is bare on the windward side. The tree resembles a banner. Encouraging for WECS, probably 12–19 mph winds.
- Class V, Partial Throwing, Clipping, and Flagging. A partially thrown tree is one

in which the trunk, as well as the branches, are bent to the lee. The trunk may be bent in a concave or convex fashion, but rises vertically near the ground and the degree of bending increases near the top of the trunk. Should be a winning area for WECS, 13–22 mph mean annual winds probable.
- Class VI, Complete Throwing and Flagging. The tree grows nearly parallel to the ground and along the path of the prevailing wind. The larger branches on the leeward side may extend beyond the tip of the trunk. More indicative of storms and danger than encouraging.
- Class VII, Carpeting or Extreme Flagging. The wind is so strong or accompanying conditions so severe (e.g., ice is present) that the tree takes the form of a shrub. Upright leaders are killed and lateral growth predominates. The crown grows across the ground like a prostrate shrub. Dangerous site for WECS. Special design probably required.

Even though no specific method exists for using *social and cultural indicators* of wind characteristics, valuable information can be inferred from human interaction with, and behavioral changes because of, the wind.

People don't normally choose to live in high-wind areas. If a town or farm is sited in an obviously sheltered location, the area surrounding the sheltered location may be windy. People often grow wind breaks on the upwind side of an area they wish to protect if they cannot otherwise avoid the windy areas.

Land cannot be cultivated for long in a windy area before it will no longer sustain crops. In regions of mixed open rangeland and cultivated fields, the range areas are more likely to be windy.

Signs along roadways often reflect wind potential. "Blowing sand" or "blowing snow" warning signs probably indicate local topographic enhancement of winds. State highway patrols can be a source of information on wind because of their continual concern with road conditions. Accidents where trucks have had their loads blown off may be indicative.

Wind damage to buildings, billboards, power lines, etc., can also be considered. Damage to structures may indicate extreme windspeeds or turbulence to be avoided, but surrounding areas could have strong but more benign wind.

Long-time residents of an area, especially those who work outdoors and cover a large territory, often can provide useful wind information. A resident may be able to say with considerable confidence that one region has more wind than another.

Even formal or informal place names, such as "Windmill Hill," "Windy Gap," or "Windy City" can provide clues (although a Maryland sign pointing to "Windmill Hill" led me neither to a hill nor any evidence of a windmill).

When all is said and done, unless potential sites are so obviously lacking in wind that they should be abandoned or so obviously rich in winds that they should be built upon without delay, *measuring and recording wind characteristics* at specific candidate sites cannot be avoided. The only questions are:

- How much detail is needed?
- How accurate do the data have to be?

- How long do data have to be collected before responsible conclusions can be made?
- How can the job get done?

The answers to these questions depend on who you are, what you want to accomplish, and the economic consequences of dependence on the data which evolve.

If you want to install a WECS for public relations, research, protest, or for the adventure, you don't need much wind data and can save the cost and time of specific wind measurements.

If you're seriously considering installing a fleet of WECS to save on fuel or energy purchases or to sell the electricity to make money, then a thorough understanding of the winds which will interact with your anticipated WECS is among the most important factors in determining economic feasibility. In this case, investing the time and money for accurate, professional, on-site wind resource assessment is essential.

Measuring, recording, retrieving, analyzing, and presenting wind characteristics is a meteorological function. The work can be performed by a professional meteorologist or by an experienced wind engineering consultant. If such people are not on your staff, you probably can find one or more in your area (if it is a known windy area) who will be willing to contract for all or any part of the job. Many WECS dealers offer basic site assessments but they have to be careful to avoid a conflict between their interest in selling their hardware and providing objective analysis of a site.

Whether the work is done in house or out, a thorough wind-resource study will include the following:

- Near continuous measurements of wind velocities and direction relative to time of day, day, month, season and year at both a reference height (usually 10 m or 30 ft) and at one or more higher points which might be indicative of WECS rotor locations.
- Notations of historical and design extremes in windspeeds, relative weather and environmental conditions (such as expectation of icing, blowing sand, or salt spray), and obvious site conditions (such as location, elevation, potential wind obstructions, and terrain features) should be made.

- All data and observations should be retrieved, processed, and analyzed.
- Data should be presented in a useful form along with conclusions and recommendations.

CONSIDERATION OF THE PUBLIC

In addition to having a buildable site with good winds, many additional factors will have to be considered and, in some cases, dealt with before proceeding with installation of WECS. The public interest, as well as the windpower developer's, must be served. In some circumstances, the law demands it. In others, common sense suggests it.

There are many positive public serving aspects and a few negatives. The fact that windpower is a plentiful and economical energy resource is no guarantee that it will be widely used. One of the clearest energy lessons of the last decade is that the technologies which a particular society relies on often bear little relation to either economic reality or human needs.

The public concerns have been grouped into three broad categories for discussion here:

- Environmental;
- Safety;
- Legal and political.

Environmental Concerns

Most environmental impacts of WECS are positive and, in fact, were the driving force behind the reemergence of windenergy conversion technology in the '70s.

On the plus side, conversion of wind to useful electricity:

- Reduces need for depletable fossil fuels such as oil and natural gas;
- Reduces dependence on foreign countries for fuels;
- Helps balance-of-payments situation by reducing imports;
- Doesn't pollute the air;
- Doesn't pollute rivers and streams;

- Doesn't require scarce water for cooling;
- Leaves no hazardous wastes to be disposed of;
- Uses free and renewable fuel which is locally available;
- Disturbs the land only minimally, usually much less than other energy generators;
- Is the most energy efficient (energy out compared to energy in) of all the available electricity technologies;
- Is generally environmentally benign.

In the neutral or innocuous category are:

- Aesthetics, which are in the eye of the beholder;
- Effect on animals, fauna, climate, bird migration patterns, and audible noise.

From the public's viewpoint, the only known potential negatives relate to potential design problems with:

- Infrasound which might cause local vibration;
- Electromagnetic radiation interference which might cause near-by television distortion.

Operational noise is often questioned if WECS are being considered for installation near people or where excess noise would otherwise be objectionable. The audible noise issue is easily disposed of by recognition that experience has consistently demonstrated that background noise caused by the wind is almost always of greater intensity than the noise of the operating WECS itself. At a reasonable distance, dictated by normal safety considerations, any noise emanating from properly functioning wind driven WECS is simply not heard. Objectionable noise can be caused by a mechanical or lubrication problem in the WECS hardware. In such cases, the malfunction should be repaired. Very small WECS generally will be more noisy than larger models because their blades rotate much faster. Even then, the noise level of properly functioning hardware should not be objectionable to anyone a safe distance from the rotating equipment. Noise from very small machines has been described as "a light swishing sound." Noise from larger machines is more of a "deeper whir."

The question of infrasound (acoustic energy below the limit of human hearing, around 15 Hz) was made real by DOE/NASA's installation of their MOD-1 (2 MW) research HAWT on a mountain near Boone, North Carolina, in 1979. The MOD-1 used rigid, as opposed to flexible, design principles which resulted in a "stiff" machine. The rotating blades initiated vibrational frequencies coincidental with the fundamental harmonic frequency of the tower structure. The resulting resonance raised the infrasonic sound to uncomfortable levels which were felt rather than heard. Area residents reported vibrations in their houses; dishes on shelves reportedly trembled. The specific solution to the MOD-1 problem was a change in the rotor's operating speed to avoid the forcing frequency. NASA's generic answer for large HAWTs was a switch to a flexible, or "soft," design philosophy which appears successful.

Therefore, neither audible noise nor infrasound is a problem with properly designed and functioning WECS hardware. Hopefully, commercially available WECS have profited from the embarrassing DOE experience and will not have similar problems.

The possibility of electromagnetic radiation interference (EMI) by improper siting of large WECS is frequently of concern. Interference may occur when signals reflected from moving rotor blades interact with the original signals, causing fluctuations in signal frequency and amplitude which degrade reception quality. Types of signals which may be affected are in the higher frequencies such as television and microwave at points where geometries favorable for interference occur among the wind turbine, transmitter, and receiver. Other factors affecting the magnitude and severity of this impact include blade area and speed, direct signal strength, and reflected signal strength relative to the direct signal.

HAWTs reportedly present a better reflecting target than VAWTs because most TV signals are horizontally polarized. Metal blades appear to have somewhat more potential for interference than non-conductive materials.

DOE/NASA and EPRI studies, and limited tests at the NASA MOD-0 site in Ohio, concluded that interference increased as the TV transmission frequency (the channel number) increased, and decreased as the distance of the television receiver from the HAWT increased. The interference radius increases with the distance of the TV transmitter from the HAWT.

Some reception difficulties may occur as far away as one-quarter mile (400 m) for low frequency (VHF) TV signals and three miles (4800 m) for higher frequency (UHF) signals.

If WECS are being considered for sites where they may be in the path of vulnerable transmissions, the project should be carefully designed in accordance with FAA and FCC standards to eliminate or reduce the potential problems. In the unlikely event that new residences are planned near WECS projects, they should utilize cable television (such as was incorporated as part of the DOE/NASA installation of a prototype MOD-0A HAWT on Block Island, Rhode Island) or use highly directional fringe antennas, install a local TV repeater, or have affected transmissions circularly polarized.

Microwave communication links also can be muddled by the scattering of secondary interference signals reflected off a rotor. Such effects can be eliminated by observing a "forbidden" zone around the microwave-link receiver.

In general, if properly sited at good wind sites in non-urban areas (as recommended elsewhere) interference with transmission signals in the air should not be a significant problem with modern WECS. Where site-specific conditions require consideration, design criteria are available.

Safety Considerations

Safety is a most important and realistic concern from the standpoints of both WECS hardware survival and danger to people and property during construction and operation. WECS are large machines with high-speed rotating blades (usually at speeds of over 100 mph) which are expected to perform in a wide variety of climatic extremes over many years. They are extremely complex and combine both mechanical and electrical features which present potential danger.

Because WECS should be installed in windy areas where there are minimal obstructions, they are obvious targets for lightning strikes which, if ungrounded, can damage the WECS itself and cause control and electrical problems.

Attempting to assemble and raise heavy, awkward equipment high into the wind is difficult at best, and potentially dangerous if performed without proper safety procedures and equipment. Established construction industry safety standards must be faithfully followed during all aspects of construction from preparing the site to final connection and energizing of electrical lines.

The essential wind itself increases potential danger during the construction process. Those winds also provide the critical design criteria for the installed WECS hardware, both in operating and parked modes. Winds can be extreme (such as in tornados or hurricanes), can be very turbulent with high wind shears and rapid changes in direction (usually with storms), and can bring windblown dirt, sand, or salt spray to clog or deteriorate important working parts. Gusting can cause control confusion during start-up or shut-down of normal WECS operations.

Ice buildup on blades, freezup of working parts, interference and obstruction by heavy snow, interference with lubrication (as well as thermal movement of parts) by extreme temperatures, and possible earthquakes are potentially hazardous to the WECS and its neighbors.

In addition to potential mechanical hazards, WECS are part of the electrical system. They have all the risks normally associated with other electrical equipment, transmission lines and substations. During construction, high voltage lines must be avoided. During intertie with the power line or the load, only skilled electricians should be involved. Once in operation, unauthorized intrusion near the WECS should be prohibited. Special care should be taken that the WECS not feed electricity into power lines when those lines are believed dead by workers. Fortunately, utility regulations and procedures are well established to assure electrical safety.

The location of the WECS may make it a potential obstruction to navigation. If near an airport or military base, adequate lighting and painting are essential. FAA regulations deal with WECS heights, distances from flight paths or landing fields, and types of warning lights.

Once installed, WECS should be considered attractive nuisances, in the same context as a swimming pool or electrical substation. Some people will want to climb into it (in at least one instance, hang their fraternity flag on it), show it off, shoot at it, or otherwise tilt at it. If the total area or site isn't secure, the dedicated WECS site must be protected both from intrusion by outsiders who could get hurt and from mechanical failure of machine components which could result in the machine toppling, parts falling, or components or ice being thrown. Barbed wire security fencing with standard high voltage warning signs have proven effective for operating safety. Certainly, WECS should not be installed in close proximity to people or damagable property (such as automobiles and buildings). If a single WECS is being installed, a reasonable "hazardous zone" from which people and valuable property should be excluded can be defined by either a nominal circle with a radius approximately six times the diameter of the WECS rotor or, if the rotor is small in comparison to its height above grade, at least twice the distance from grade to the highest blade tip (when the blade is vertical). Although data are scarce on actual component throws during WECS failure, a blade thrown from the Smith-Putnam 1.25 MW HAWT in 1945 reportedly traveled 750 ft. DOE/NASA estimates that their MOD-0A 200 kW 62.5 ft long blades could be thrown as far as 500 ft. Small parts from the Alcoa/Southern California Edison 123 ft by 82 ft research VAWT (which collapsed in San Gorgonio Pass, California) were found over 400 ft from the base.

In addition to making sure that safety is an important part of design, construction, and WECS operations, the WECS facility should be adequately insured, either by addition to existing insurance coverage or by establishing project-specific insurance. Insurance should cover potential liabilities as well as repair or replacement of WECS hardware.

Legal and Political Considerations

Added to practical environment and safety attention, legal and political realities will be important (in some cases, obstacles) in developing a WECS project.

Land use considerations are a combination of legal and political factors which are often governed by state or local zoning laws, although non-urban areas (where WECS are most likely) do not normally have stringent local regulations. Relative restrictions include limitations on height, setback, land use, and in some cases aesthetics. The zoning process is highly decentralized, discretionary, and political in nature. It is very dependent on local interests and attitudes. Most zoning ordinances do not cover WECS specifically and very few deal with "wind rights." The local situation should be fully researched before committing to a WECS project.

"Wind rights" or "negative easements" are relatively new issues and can be important with single units or small projects where large areas of land are not under the control of the developer. The idea is to insure that a neighboring landowner does not build anything that obstructs flow of the wind to WECS rotors. In some cases, such access can be purchased under direct contract with the neighbor. The first statewide legislation intended to clarify and deal with the potential issue was passed by the Oregon legislature in the early '80s and experience is very limited. In the wide-open spaces, this is probably a non-issue.

Building codes may be in effect at some potential WECS sites. If so, the specific requirements must be met before a building permit can be issued. Like zoning, codes are customized and administered locally and should be researched on a site-specific basis. In general, codes deal with building construction and have strong emphasis on hardware and safety. WECS are not specifically included. They'll probably be treated like a building or free-standing structure, with emphasis on structural capability, foundations and attachments to the ground, and electrical wiring and connections. Often, there will be a requirement that a state-licensed professional engineer approve the site-specific design. This may require involvement of local civil and electrical engineers, a good idea even if not specifically required by law. By the mid-'80s, consensus standards for WECS will be established and will be covered in local codes by reference to ASTM and ASME standards and recognized national model codes. Until then, common sense will have to prevail.

Zoning and building codes are strongest in urban areas and are heavily oriented toward residential, commercial, and industrial land use and construction standards. Public land generally is not subject to the local restrictions. In their place, public agencies have their own regulations which are often more difficult to work with. Fortunately, cooperative efforts between state and federal agencies were initiated in 1981 to deal with many of the artificial and arbitrary regulations and to "clear" environmental impact statements and modify land use requirements in known windy areas.

PROJECT PLANNING AND ANALYSIS

If you've identified a buildable site with good winds, and believe that WECS can be installed within the practical, legal and political constraints, the next step is to fully identify and evaluate the potential project. An overall development strategy, with various scenarios to anticipate imposed or unforeseen options which may evolve, should be thought through. What do you want to accomplish? Do you need technical or financial help? Should you take on a partner? Hire a consultant? Do you need a detailed study of the project before proceeding? Or are you going to build regardless of what a study might show?

The site is the most important asset in a WECS project. Do you own or control it? Can it be developed? Can delivery trucks, cranes, and workers get to it? Is it served by a power line or will one have to be built to transmit the WECS power to where it has value? Is there enough acreage to site WECS without one interfering with another? Are there current or anticipated obstructions to wind access? What are the characteristics of the wind? Intensity? Direction? Time-of-day and days-of-year availability? Extremes expected? Accuracy of wind data? Is on-site meteorological equipment needed? How about the ground? Will expensive excavation and grading be required? Will it support foundations and anchors? Will WECS be aesthetically acceptable additions to the landscape?

The value of WECS electricity depends on what you intend to do with it. Are you going to utilize all the energy directly yourself? If so, what is the value of the purchased energy or fuel to be displaced? Does time-of-day or day-of-year affect that value? What is the correlation of direct energy needs with the expected availability of the wind driven WECS energy? Can peak WECS output always be utilized? Does it coincide with base demand? Will wind availability allow reduction of peak power demand? How much energy from the WECS can you really use? What can you do with it if you can't use it when wind availability schedules it? On-site storage? Dump it into the utility grid? If you're going to sell all the WECS generated electricity, analysis will be simpler. What will your sales price be? Now? In the future? Based on your customer's peak and minimum loads? Based on time-of-day or day-of-year? Or fuel being displaced? If you plan to use some output directly and sell excess, the analysis will get complicated. All the variables of direct use will combine with all the variables of total sale.

The site conditions and volume of electricity used or sold will combine to suggest the size of WECS project which can be justified, limited by annual energy (kWh), total power (kW), or number of machines.

When you have a pretty good idea of the size of project you want to consider, you can get at issues of size and type of WECS hardware, noneconomic risks and benefits associated with the project, and type of project ownership or structure which can be most advantageous. Finally, a thorough economic analysis can be made and feasibility, or lack of it, determined.

The WECS hardware questions relate to big versus small, many versus few, HAWT versus VAWT, availability, reliability, operating and maintenance requirements, and warranties and services available. Electrical substations or transformers, interconnecting power lines, controls and anemometry, instrumentation and metering, power conditioning or storage equipment, and lightning and electrical grounding, as well as roads and fencing will be required. The total system must be considered. How many dollars have to go in to get how much useful energy out over a period of years? That's the measure of efficiency for a WECS installation.

Noneconomic benefits of a WECS project should have some weight in comparison to the development's risks. Although difficult to quantify, there may be public relations, political, prestige, and societal advantages to help cover the risks and costs of pioneering or innovating. You also may value the degree of independence that converting the locally available wind into useful energy provides. Hopefully, all those benefits together will balance potential risks inherent with WECS hardware, WECS politics, and the wind itself.

Modern WECS hardware is relatively new. Will it last? Will it perform as predicted? Will it operate when needed without unreasonable repair or maintenance?

WECS politics seem to be more vague than the wind itself. Will tax incentives remain? Will PURPA be favorably implemented? Will public utility commissions stay friendly to wind? Will people continue to support locally available benign energy sources?

The wind itself is surprisingly predictible over time. But will it be there when needed most? Will there occasionally be wind droughts? Will something grow or be constructed that obstructs wind flow to the WECS rotors?

The type of ownership of the WECS project can have an important influence on its economic viability as well as its freedom of operation. If you're not already a utility, should you become one to fully exploit your wind resource? A cooperative? A nonprofit or community-owned facility? You might be eligible for low interest loans. If you're a taxpayer, is single proprietorship or a simple partnership your best way to structure the project for maximum after-tax benefits? Or should a limited partnership be established? How about a R&D partnership? Or a special purpose corporation? Perhaps in joint venture with a utility if you're not one. Or another utility if you are. Do you want someone else to own the WECS and sell you the electricity? Or lease you the machines for operation? Do you want to build, sell and lease back the hardware? The options are limitless and depend on what you're trying to accomplish and how much you want (or can afford) to invest. Others are available to pick up any portion of an economically attractive WECS package that you don't want to exploit yourself.

Even regulated investor-owned utilities have the option of financially participating, up to 50 percent, in non-utility windfarms or small power-producing facilities to take advantage of the unregulated climate and incentives not available to them as utilities. Those same IOUs can simply purchase electricity from third party generators, can lease WECS facilities from others and operate and maintain them, or can provide "in-kind" facilities (such as land, power lines, transformers, substations, etc.) or services (such as removal of obstacles, land surveying, site engineering, assistance with permits and environmental impact statements, grid-integration help, etc.) in an effort to keep investments by third parties, and the resultant selling price to them, as low as possible. The IOUs can, of course, directly purchase WECS for integration into their generating system.

Unregulated power producers or users (federal, state, and municipal utilities, rural electric cooperatives, and other special purpose public agencies, nonprofits, or coops) enjoy financial incentives in the form of low interest money (bond issues, REA loans, direct taxpayer allocations, etc.) because of their status, but can't take direct advantage of special WECS tax credits. They can, however, cooperate with third party entrepreneurs to buy electricity or lease WECS and to provide resources (such as listed for the IOUs) to help keep the costs down. Being unregulated, this class of potential WECS investor has more freedom of action and has been willing to take action perceived by them to be beneficial to their constituencies.

Profit-oriented, tax-paying entities can benefit from the special WECS incentives, but have to fit WECS into overall business operations. An investment in WECS may be perceived as simply prepaying electric bills (which are expensed and essentially reduced by the tax rate of the business). Most businesses don't regard themselves as being in the power business and many believe that their local utility will always take care of them. They usually will want to subject potential WECS investments to the same type of financial goals as other opportunities competing for available funds. Therefore, only projects where economic rewards appear especially high will result in installation of WECS for other than public relations or technological research reasons.

Special purpose entrepreneurs have the most freedom of action and can structure themselves to take full advantage of all incentives and cooperation available and can assign out most of the risks of pioneering innovation. Most have chosen to structure as limited partnerships (in the classic mold of real estate and oil and gas tax shelters). Some work their magic as single proprietors, cooperatives, corporations, or not-for-profit agencies. Although initial deals have been put together to own WECS hardware and sell the energy to utilities, there is no reason why the hardware or project can't be leased to others, or why projects can't be built on speculation and sold off to utilities or others for quick capital gain.

Determining economic feasibility of a WECS project can be extremely complex and is always site-specific. Before investing time and money in a full-scale feasibility study, consider the following:

- Ignore all comparisons of investments ($) for power (kW). With intermittent wind as fuel, power (kW) output of WECS will vary substantially from instant to instant. However, the investment doesn't vary.
- Don't try to outguess the WECS producer on power or energy projections. Get the producer's best available power curves, or power vs. windspeed tables, for each machine being considered. Use those power data to project energy (kWh) based on actual or best guess winds for the specific site. Don't expect more accuracy in the energy projections than is available in the wind data.
- If actual wind data are not available for the project site, but there is reasonable confidence in average windspeeds, use Rayleigh distribution of the windspeed frequencies to plot energy expectations based on WECS power at each windspeed. Most

producers project energy for "typical" wind regimes. Adjust them to site-specific conditions.

• Estimate the installed cost of WECS hardware, including anticipated off-site (such as engineering, management, shipping, etc.) and site improvement (such as transformers, roads, fences, etc.) costs. Producers, or their representatives, should be able to provide a reasonable range of costs if the location and time of construction is known.

• Assign value to the WECS generated energy, based either on avoided electricity purchases or sale of the electricity, or an expected mix of the two. Multiply the expected WECS generated energy (kWh) by the energy value ($/kWh) to establish a current monthly or yearly value of each WECS being considered.

From the preceeding, a first look at simplistic economics (see Figure 5-4) can evolve useful results, such as:

Because wind conditions vary almost continuously, analyzing the cost of WECS based on investment per unit of power (kW) is meaningless. Alternatives for early analysis include:

• Cost of annual energy ($/AkWh);
• Annual energy per dollar invested (AkWh/$);
• Time to pay back investment (Years).

Assumptions:

• WECS will generate one million kWh annually (AkWh);
• WECS will cost $500 thousand installed ($);
• Current value of energy = $.08/kWh.

Analysis:

$/AkWh   = $500,000/1,000,000 AkWh = $.50/AkWh
AkWh/$ = 1,000,000 AkWh/$500,000 = 2 AkWh/$
payback years = $500,000/1,000,000 × $.08/kwh = 6.25 Years
            or $.50/AkWh/$.08/kWh = 6.25 Years

Figure 5-4. Simplistic economic analysis.

- Cost of annual energy ($/AkWh), by dividing total estimated cost by total annual energy expected from the WECS;
- Annual energy per dollar invested (AkWh/$), by dividing total annual energy expected by the total cost of the WECS project;
- Time to pay back the investment (years), by dividing the cost of annual energy ($/AkWh) by the current value of the energy ($/kWh) or dividing total project cost ($) by the yearly value ($/year).

Based on knowledge of local energy or fuel availability and costs, now and in the future, as well as available federal and state tax credits and any other incentives, those results can be quickly put into perspective. A judgment can be made as to whether the WECS project has any hope of economic feasibility within your criteria for favorable consideration. If not, the project can be abandoned or built for non-economic reasons. If it is attractive, additional analysis is advisable.

An intermediate analysis, which might be helpful, is to consider cost of energy (COE) expected from WECS. An easy-to-use formula was established early in the U.S. Government WECS Research Program to compare one WECS to another or a proposed WECS to a cost goal. COE is based on just five variables:

- Installed cost (IC) of the WECS, or total investment;
- Annualized cost factor (ACF, percent of investment) for owning the WECS; percentages range from eight to 22, depending on owner and local conditions;
- Annual operating and maintenance (AO&M, in either annual dollars or as a percentage—commonly 2-4 percent—of the initial cost) budget;
- Tax credits (TC) available, federal plus state;
- Annual energy output (AkWh) expected from the WECS in a given wind regime.

The formula is:

$$\text{COE (\$/kWh)} = \frac{(IC - TC) \times ACF + AO\&M}{AkWh}.$$

Although the annualized cost factor will vary with each owner, typical percentages utilized are:

Government agencies, 8–10%
Publicly-owned utilities, 9–13%
Rural electric cooperatives, 12–15%
Small businesses, 15%
Big businesses, 15–18%
Investor-owned utilities, 18–22%

Those percentages were intended to reflect interest rates, expected returns on investments, depreciation, taxes, insurance, and other economic costs and benefits to typical buyers. Figure 5-5 provides an

DETERMINING COST OF WECS ENERGY (COE)

Cost of energy (COE) generated by a WECS depends on the total initial installed cost (IC), deductions for available tax credits (TC), an estimate of annualized operating and maintaining costs (AO&M), an estimate of annual energy output (AkWh) based on wind conditions, and an annualizing cost factor (ACF) which reflects expected cost of money and other ownership factors.

Assumptions:

AkWh	= 1,000,000 kWh
IC	= $500,000
TC	= 25% or $125,000 (to taxpayers only)
AO&M	= 2% of IC or $10,000
ACF:	Government agency 10%
	Publicly-owned utility 12%
	REC and small business 15%
	IOU and big business 20%

Analysis:

OWNER	ANNUAL COST	COE
Government agency	$ 60,000	$.060/kWh
Publicly-owned utility	$ 72,500	$.072/kWh
REC	$ 85,000	$.085/kWh
Small business	$ 66,200	$.066/kWh
IOU	$110,000	$.110/kWh
Big Business	$ 85,000	$.085/kWh

Figure 5-5. Cost of energy (COE) illustrations.

illustration of COE. If the COE can be determined with confidence, that cost should be compared with current and projected costs of energy purchases displaced or revenue from WECS energy sales. If COE is lower than current value, the investment is obviously attractive. If higher, estimates of inflation's effect on future energy costs or selling prices will have to be made for further comparison.

If the project still appears to be favorable, a detailed *feasibility study,* including full economic analysis, should be prepared. The accuracy and quality can't be expected to be any better than the accuracy and completeness of the data used to prepare it. If project costs, site-specific wind data, and value of energy production are not known, further analysis can only be theoretical or provide rough approximations of financial requirements and potential rewards. That's okay if the reliability of output data is kept in perspective and utilized accordingly.

For bankable results, a detailed site-specific analysis must be made. The project must be designed with enough detail that accurate cost budgets and construction schedules can be determined. Reliable wind data must be accumulated for at least a year at locations relative to the planned WECS rotor locations. Accurate value of the energy must be determined in conjunction with utility rate structures which ultimattely can be turned into contractual obligations. Project schedules which evolve will affect costs and revenues and will also impact considerations of available incentives and depreciation options. Finally, the choice of legal structure of the WECS project must be made before the economic pro forma can be created and analyzed.

Project strategy should anticipate unfavorable as well as favorable events and provide contingencies for them. In attempting to strategize a successful WECS project, consider the points in Figure 5-6. Call on the best resources available for project planning, site design, estimating and analyzing results. If you have all the necessary skills in house, fine. If professionals are needed to complement internal capabilities, look for people or firms experienced in project development. If experience with WECS is not available, experience with real estate development is directly relative to most considerations.

An entrepreneurial project manager should pull everything together. Legal, financial, accounting, tax, construction, engineering, meteorology, procurement and estimating specialists will be needed for

WECS PROJECT STRATEGY CONSIDERATIONS

1. Equity, debt and leverage
   A. Minimize equity and cash requirements.
   B. Borrow as much as possible, for as long as possible.
   C. Borrow at lowest possible interest rate; use subsidies if available.
   D. Secure grants if available.
   E. Weigh nonrecourse financing vs. tax considerations.
2. Tax credits and depreciation
   A. Utilize federal investment tax credits (10 percent) and energy tax credits (15 percent) directly, if possible, or indirectly, if necessary.
   B. Utilize state tax credits where available. 1982 examples include:
      —California, 25 percent.
      —Hawaii, 10 percent.
      —Oregon, 35 percent (10, 10, 5, 5, and 5 over five years).
   C. Double dipping is not allowed:
      —No tax credits on federal grants.
      —No tax credits on federally subsidized loans.
      —Debt recourse definition is a factor.
   D. Utilize 1981 ERTA accelerated cost recovery system (ACRS) five-year depreciation:

| RECOVERY YEAR | FOR PROPERTY PLACED IN SERVICE IN: | | |
	1981–1984	1985	POST-1985
1	15%	18%	20%
2	22%	33%	32%
3	21%	25%	24%
4	21%	16%	16%
5	21%	8%	8%

3. Management and planning
   A. Bring projects on line as near the end of the year as possible to maximize tax credit and depreciation effects.
   B. Assign as many risks as possible to others.
   C. Accumulate start-up expenses for later depreciation.
   D. Minimize outflow of cash, consider escalation.
   E. Maximize inflow of cash, consider inflation.
   F. Realistically estimate production of WECS.
   G. Protect cost estimates (contingencies, provision for O&M, etc.)
   H. Work with "world class" suppliers and associates to maximize credibility.
   I. Think in after-tax terms (paper losses, positive cash flow).

Figure 5-6. Strategy checklist for WECS project.

professional input. Some of the factors important to a thorough feasibility study are shown in Figure 5-7 as a checklist. After the bulk of that data is available, a meaningful economic analysis and illustrative financial proforma can be prepared.

Figures 5-8 and 5-9 show examples of economic analysis based on two different WECS installed in varying wind regimes.

- Project description
  - location and site plan
  - type and quantity of WECS hardware
  - wind resource description
  - energy projections and schedules
  - objectives and participants
  - implementation schedule
  - assumptions vs. hard data
- Construction cost estimate
  - land
  - site preparation
  - WECS purchase and delivery
  - assembly and erection of hardware
  - start-up and testing
  - interest and taxes during construction
- Start-up and annual expenses
  - structuring of project
  - management, administration and marketing
  - professional services
  - property taxes
  - insurance
  - operating and maintenance
  - repayment of debt
  - interest on debt
  - depreciation
  - expense escalation assumptions

- Revenue or savings estimate
  - energy value
    initial
    escalation for future years
    average vs. time of generation
  - tax credits
    federal
    state
  - tax deductions
    depreciation
    interest
    taxes
- Financial analysis
  - equity
    amount
    expected returns
  - debt
    amount
    life
    interest rate
    terms
  - cash flow
  - taxable income (profit or loss)
    amount
    applicable tax rates
  - after tax returns
    amount
    rates
  - estimated future value of project
    sale
    salvage
    rehabilitation

Figure 5-7. WECS feasibility checklist.

WECS PURCHASE EXAMPLE: FINANCIAL ASSUMPTIONS
(Private Investor Eligible for Federal Tax Credits)

- 500 kW WECS, one machine or several
- Installed cost = $1,000,000 (includes site costs); or $2,000/kW
- Annual O&M = 2% of capital cost
- Annual property tax = 1% of capital cost
- Annual insurance cost = 0.5% of capital cost
- Expense escalation rate = 7%
- Annual energy production = 2,000,000 kWh
- Total tax credit = 25% of investment

- Investors' tax bracket = 50%
- Depreciation period = 5 years (ACRS)
- Initial energy value = 7¢/kWh
- Annual energy escalation rate = 12%
- Equity = $250,000
- Debt = $750,000
- Length of loan = 15 years
- Loan interest rate = 18%

PROFORMA REVENUE, EXPENSES AND RETURNS

YEAR	REVENUE	EXPENSES	DEBT SERVICE	INTEREST	DEPRECIATION	TAX CREDIT	CASH FLOW	TAXABLE INCOME	AFTER TAX RETURN
1	$140,000	$35,000	$147,302	$135,000	$150,000	$250,000	$207,698	-$180,000	$297,698
2	$156,800	$37,450	$147,302	$132,786	$220,000	0	-$ 27,952	-$233,436	$ 88,766
3	$175,616	$40,071	$147,302	$130,173	$210,000	0	-$ 11,757	-$204,628	$ 90,557
4	$196,690	$42,876	$147,302	$127,089	$210,000	0	$ 6,511	-$183,276	$ 98,149
5	$220,292	$45,877	$147,302	$123,451	$210,000	0	$ 27,112	-$159,036	$106,631
6	$246,728	$49,089	$147,302	$119,158	0	0	$ 50,336	$ 78,480	$ 11,095
7	$276,335	$52,525	$147,302	$114,092	0	0	$ 76,507	$109,717	$ 21,648
8	$309,495	$56,202	$147,302	$108,114	0	0	$105,990	$145,179	$ 33,401
9	$346,634	$60,136	$147,302	$101,060	0	0	$139,196	$185,438	$ 46,476
10	$338,230	$64,346	$147,302	$ 92,736	0	0	$176,582	$231,148	$ 61,009
						TOTAL:	$750,223	-$210,415	$855,430

After tax internal rate of return = 68%

Figure 5-8. Financial assumptions and illustrative proforma for purchase and installation of 500 kW WECS for $1,000,000. Courtesy of Mike Lotker, The Synectics Group, Inc.

193

VAWTPOWER 185 PURCHASE EXAMPLE: FINANCIAL ASSUMPTIONS

- 185 kW VAWT, One machine
- Installed cost = $150,000 (Includes site costs)
- Annual O&M = 2% of capital cost
- Annual property tax = 1% of capital cost
- Annual insurance cost = 0.5% of capital cost
- Expense escalation rate = 7%
- Annual energy production = 300,000 kWh
- Investment tax credit = 25%

- Investors' tax bracket = 50%
- Depreciation period = 5 years (ACRS)
- Initial energy value = $.06/kWh
- Annual energy escalation rate = 12%
- Length of loan = 15 years
- Loan interest rate = 18%
- Amount of debt = $112,500
- Amount of equity = $37,500

PROFORMA VAWTPOWER 185 FINANCIAL INVESTMENT CALCULATIONS

YEAR	REVENUE	EXPENSES	DEBT SERVICE	INTEREST	DEPRECIATION	TAX CREDIT	CASH FLOW	TAXABLE INCOME	AFTER TAX RETURN
0	Initial Investment		0	0	0	0	-$37,500	0	0
1	$18,000	$5,250	$22,095	$20,250	$22,500	$37,500	$28,155	-$30,000	$43,155
2	$20,160	$5,618	$22,095	$19,980	$33,000	0	-$ 7,553	-$38,438	$11,666
3	$22,579	$6,011	$22,095	$19,526	$31,500	0	-$ 5,527	-$34,458	$11,702
4	$25,289	$6,431	$22,095	$19,064	$31,500	0	-$ 3,238	-$31,707	$12,616
5	$28,323	$6,882	$22,095	$18,518	$31,500	0	-$ 653	-$28,576	$13,635
6	$31,722	$7,363	$22,095	$17,874	0	0	$ 2,264	$ 6,485	-$ 979
7	$35,529	$7,879	$22,095	$17,114	0	0	$ 5,555	$10,536	$ 287
8	$39,792	$8,430	$22,095	$16,218	0	0	$ 9,267	$15,144	$ 1,695
9	$44,567	$9,020	$22,095	$15,160	0	0	$13,452	$20,387	$ 3,258
10	$49,915	$9,652	$22,095	$13,912	0	0	$18,169	$26,352	$ 4,993
						TOTAL:	$22,391	-$84,275	$102,028

Ten year after tax internal rate of return = 60%

Figure 5-9. Financial assumptions and illustrative proforma for purchase, installation, and operation of specific 185 kW VAWT. Prepared by Accountant Gary Smith for Forecast Industries, Inc.

Comparison of the financial results shows the dramatic effect of tax credits, depreciation and interest rates on the value of the WECS.

WECS PROJECT DEVELOPMENT

Once you've found justification to proceed, development of a WECS project will be similar to any other real estate or construction project. Many skills required are the same as those needed for the feasibility study, but objectives and required management skills shift from planning to execution. Effective scheduling and budgeting are essential, but getting WECS up and operating successfully will require project leadership and construction management equivalent to that needed for any capital expenditure. How it gets done depends on available resources and experience.

You may want to do the whole job with your own staff and construction people, or it might be advantageous to contract all work to specialists. Probably a combination of the two, which combines the best of both, will provide superior results. The choice of participants and project structure will also depend on broader objectives and future plans. If you're only interested in one WECS project, it probably would be effective to buy a turnkey installation, perhaps complete with an operating, maintenance, and service contract. If the project is considered a prototype for additional or ongoing operations, you may want to build up an internal capability to at least manage the total process, so that the most effective development team can be put together for each future effort.

The specific development process will vary from project to project but should start with participation in the feasibility study and its conclusions. Obviously, if the executors can't perform within the ground rules strategized by the planners, the results will differ from the plan and change economic requirements and outcomes.

An effective project plan provides for an orderly transition from the feasibility study to execution, starting with confirmation of assumptions and fleshing out of preliminary design, engineering and scheduling. The general development steps can be categorized as follows:

- Formalize an action plan;
- Legitimize structure and assumptions;
- Complete the site evaluation;
- Design and engineer the site;
- Execute contract documents;
- Prepare the site;
- Assemble and install WECS hardware;
- Operate and maintain the facility;
- Terminate and dispose of the project.

In *formalizing the plan* of action, the first step is to put a competent leader in charge and then define and organize the project team. Many of the participants in the feasibility study should continue, but the emphasis must be changed from planning and analysis to getting things done. Objectives and decision points must be clearly established. Budgets, schedules, and administrative procedures should support the objectives. All work tasks must be identified and responsibility and authority for each action assigned. As soon as the plan is ready, it should be brought to life.

An obvious early action is to *make sure everything is legal and achievable.* If a new business entity was anticipated, set it up. If assumptions were made, verify or modify them. If the site is not clearly owned or controlled, bring it under control. If outside influences are expected, identify and deal with each. Will a building permit be needed? How about approval by a land use agency such as your local planning or zoning board? Or the state? Will the Feds have to approve anything through FAA, FCC or other agencies? Will licenses be required for things such as contract documents, on-site work or operations? Is financing in hand, or does money still have to be raised? Make sure expectations are real. Are environmental or political issues unresolved or likely to emerge? Resolve or preempt them.

If you're not a utility, is your utility on the team? If not, get them on. If you are a utility, are you clean with your regulators and constituencies? If not, get clean. If the buying and selling prices of future electricity are not legally clear, make them clear. It is time to replace assumptions and wishful thinking with reality. If the project turns sour, stop it quickly before the cost goes up.

A thorough *evaluation of the site and its winds* must be completed and data quantified and qualified. Make sure wind is measured at the proper locations and that data are accumulated over a long enough time to be meaningful. If the wind resource was overestimated, reconsider the entire project. Is the wind going to reach the WECS in the future? Or is something growing or being built which may interfere or obstruct? If necessary, make a deal for rights to the wind with appropriate neighbors. Are there restrictions and covenants on your land that will have to be modified or designed around? Height limitations? Noise restrictions? Easements? If so, do it. Are codes, standards, land use requirements and procedures understood and being followed? What are the site conditions which will dictate design and engineering? Topography? Soil bearing? Digability? Snow and ice? Expectation of lightning? Temperature extremes? Windblown contaminants or corrosives? Earthquake conditions? Extreme wind conditions? Make sure they are known and dealt with. Test the soil and survey the land if necessary. Improve measuring and recording instruments and extend time if necessary. Locate important interface points. Access to power line? Access to roads for hardware delivery? Potential safety or security hazards? Know your site conditions before time and money is invested in modifying them.

Site design and project engineering can proceed when the site conditions are known. Design should combine best land use with best wind use. In addition to considerations of construction and operations, design should consider aesthetics, security, safety, phasing of construction for incremental start-ups, avoidance of EMI, and control and dispatching strategies for the total facility. Should the project utilize one or a few large WECS, or more smaller units? Will HAWTs best relate to local conditions? Or VAWTs? Or a mixture of the two types? Perhaps HAWTs high and VAWTs low for best power density. Can simple mounding be utilized to shield the WECS from neighbors or thoroughfares? Or raise one WECS above the obstruction of another? Before zeroing in on specific WECS choices, make sure they are available within project objectives and constraints (cost, delivery schedule, reliability, conditions of sale, etc.). Don't design a project around hardware that won't be usable. Even preliminary civil and electrical engineering will require knowledge of the specific hardware to be utilized. It's hard decision time. And back to the legalities.

When the choice of WECS hardware suppliers is made, complete *contract documents* can be prepared and executed. A purchase agreement should be negotiated with the WECS producer to identify all aspects of hardware costs, delivery schedule, terms and conditions of payment, warranties, and services available. Will the WECS supplier provide a turnkey installation? Will the supplier install on foundations prepared by others? Are site preparation guidelines available? How about an assembly and erection manual? When the WECS supplier role is clarified, all remaining project work can be assigned and contract documents prepared. Site development and preparation is one distinct "conventional" phase, probably best defined by a combination of civil, electrical and construction engineers. If the WECS hardware isn't purchased on an installed and operating basis, the remaining engineering phase is preparation of contract documents for assembly and installation of the WECS hardware and special-purpose ancillary equipment. When all drawings, specifications and general conditions are complete, competitive bids can be solicited or work can be negotiated with qualified potential contractors if the work isn't to be performed in-house. Contracts or internal work orders can then be executed.

Site preparation should proceed prior to delivery of WECS hardware. Skills and equipment needed are different from those required for assembly and installation and can best be performed by a separate contractor or work team. Site preparation includes grading, excavation, roads, fences, foundations, pads, anchors, substations or transformers, electric service lines, underground wiring and lightning protection, conduits, meters and instrumentation, anemometry, and any special safety, security or communications provisions. Make sure insurance and safety programs are implemented.

When the site is ready and hardware is delivered, the *assembly and installation* phase of construction can begin. Provision for on-site unloading, storage, protection, and construction logistics should be made. Space for on-the-ground assembly and crane or special equipment maneuvering should be allocated. Installation of towers, controls, and ancillary equipment can proceed while any necessary assembly of rotors or nacelles evolves. Effective scheduling and construction management will minimize construction time and cost. When assembly is complete, the WECS hardware can be erected or installed and the process of hooking up, checking out, starting up, debugging, and responsibly turning over to operating personnel can occur. When the WECS facility is effectively operating, the development project can be concluded.

Operation and maintenance of the WECS facility includes all activities necessary to keep the equipment up and operating at peak efficiency to maximize electricity production. It includes consideration of warranties, service contracts, insurance, maintenance, and performance. Is the machine doing what was expected of it? Quality of electricity? Power versus windspeed? Cut-in and shut-down windspeeds? Response times? Are the proper credits being given for electricity generated? Is there an opportunity to improve anything?

Terminating and disposing of the project can have important economic ramifications. To take full advantage of tax credits and accelerated depreciation, the facility must be owned for at least five years. Should new ownership be considered at that time? How about a sale and leaseback? Without tax benefits, when will the returns on investment be attractive to others? When will O&M or repair and replacement costs exceed their value? Will salvage value of the total WECS or any of its recyclable components or materials make the hardware more valuable in the future? When you've proved reliability, will the WECS provide capital gains potential? Cash flows and returns should be monitored relative to potential value of the WECS to others. There probably will be an optimum time and means to terminate and dispose of the project.

CONCLUSION

Although experience with modern WECS is limited, properly conceived and developed WECS projects can be economically attractive now. Rapidly rising energy and fuel costs present the opportunity for effective alternatives. Expected public benefits of utilizing nonpolluting, renewable, and locally available winds as fuel for electricity have brought forth liberal financial incentives. If you have a windy, buildable site, you have a valuable resource which should be favorably considered for WECS development. The hardware is available and the environment is supportive. Adventure and risk associated with innovation and pioneering are present. If needed, help is available. The rest is up to you. Enjoy.

Appendix A
WECS Historical Evolution

Although use of wind power dates back to before Christ, electricity generating wind turbines were not invented or demonstrated until the 1860s when Moses G. Farmer built a small unit to power an incandescent light bulb.

The early days of windpower have been traced back to crude sailboats and grain grinders in China, hundreds of years before the birth of Christ. The panemones of ancient Persia were first reported in use in 200 B.C. and those vertical axis rotors quickly spread throughout the Islamic world. (See Figure A-1.)

Later, probably in the Tenth Century, those simple vertical axis machines were transformed into horizontal axis configurations with radiating sails and were widely utilized all around the Mediterranean Sea for water pumping and grain grinding. (See Figure A-2.) The most famous illustration of this type of windmill can still be seen on the "Plateau of Ten Thousand Windmills" on the Island of Crete at Lasithi. (See Figure A-3.) An excellent replica of the Greek jib sail mills, as well as other historic windmills, can be seen at Windfarm Museum on Martha's Vineyard, Massachusetts. (See Figure A-4.)

The returning Crusaders brought back the idea of windmills from the Near East to northern Europe. The first four-armed, cross-shaped European windmills were post mills in which the entire millhouse could be turned into the wind on a huge post. (See Figure A-5.) Artwork of the period indicates the first post mills were in place in the Twelfth Century and at least one artist pictured Don Quixote tilting at a post mill rather than the more common portrayal of the "Dutch windmill" which came later. (See Figure A-6.) A Lufthansa Airline magazine (*Jet Tales*, February, 1981) article refers to the early post mill as the "German mill" or "trestle windmill" where the whole millhouse is turned in the direction of the wind as opposed to the "tower windmill" which has a solid structure and only the roof cap and its wind wheel turn to face the wind. (See Figure A-7.) It is this latter design that evolved into the Dutch version that most people visualize when they think of a windmill. Both types of horizontal axis windmills soon became very important in medieval Europe, first for grinding grain and later for sawing wood, making paper, and draining water from low-lying farmland.

Figure A-1. Early vertical axis windmill.

Regardless of who might have been first, by the Fourteenth Century the Dutch had clearly taken the lead in windmill technology and construction, and were using them extensively for draining the lakes and marshes of the Rhine River delta. Sophisticated versions of this technology were found throughout Europe by the Fifteenth Century. Windmills, along with waterwheels, became the dominant power which greatly increased the productivity of agrarian economies.

By the Sixteenth Century, the Dutch had made many improvements to the basic windmills and the generic "Dutch" windmills had taken shape. (See Figure A-8.) Large industrial mills could deliver up to 90 hp of mechanical power in good winds. The first windpowered oil mill was built in 1582. The first paper mill was powered by wind in 1586 to meet the demands that came with the in-

Figure A-2. Early horizontal axis windmill.

vention of the printing press. By the end of the century, the Dutch were using windpowered sawmills to process timber imported from the Baltic.

Dutch windmill fame grew when the below-sea-level wetlands area of Beemster Polder was drained between 1608 and 1612 by 26 windmills with capacities of up to 50 hp each. Later, the Schermer Polder was drained in four years by a combination of 14 windmills pumping water at a rate of a thousand cubic meters per minute into a storage basin and another 36 mills pumping from that basin into a canal which emptied into the North Sea. (See Figures A-9 and A-10.) Overall, an estimated twelve thousand windmills were in use in Seventeenth Century Holland. Approximately the same number, with capacities of 10–20 hp, were in service in England.

All went well for the windmill industry in Europe until the introduction of the steam engine and the resulting Industrial Revolution. By the middle of the Nineteenth Century, less than ten thousand of Holland's windmills remained and, by the turn of the Twentieth Century, only about 2500 were still in operation. Today, fewer than a thousand are still in working condition.

Denmark made further improvements in mechanical windmills and eventually fueled almost a quarter of its industrial energy needs with the wind. The peak Danish windpower capacity was estimated to be the equivalent of 200 MW by the end of the Nineteenth Century.

Figure A-3. Valley of Lasithi with 10,000 windmills for pumping water.

The roots of modern wind turbines go back to the mid-Nineteenth Century in the United States. The colonists had brought European "post" and "Dutch" windmills with them in the 1700s and few improvements were made until 1854 when the "American multi-blade windpumper" was developed in response to water needs of the railroads and the growing western frontier. (See Figure A-11.)

In the early 1850s, John Burnham, described as a roving "pump doctor," challenged Daniel Halladay, a small machine shop operator in Ellington, Connecticut, to develop a better windmill than the available European-style machines. When Halladay's efforts appeared promising, he and Burnham opened a factory

Figure A-4. Greek jib sail mill replica at Windfarm Museum.

in South Covington in 1854 to produce his multi-vane windpump. Their first customer was Dr. H.A. Grant in nearby Enfield. However, the new company had trouble getting started in conservative New England. Burnham went west to Chicago, "the windy city," to enlist entreprenurial investors and sign up initial sales to the railroads so they could fill their trackside water tanks. The new company became the United States Wind Engine and Pump Company. When it outgrew its Connecticut factory, it moved to the suburban Chicago town of Batavia. From that Chicago base, the products grew in quantity and size (up to 60 ft diameter wind wheels) and expanded with branches in Boston, Omaha, and Fort Worth. Competitors sprang up in the Chicago area and it became known as "the windmill city," just as the Detroit area became known as "the motor city." The "Halladay Standard" windpumper was manufactured in the United States (and under license in Germany) until 1929.

Another prominent pioneering effort was by Rev. Leonard R. Wheeler, whose work led to an 1867 patent covering a "solid," rather than Halladay's "sectional," fan. Rev. Wheeler and his son William built their Eclipse Windmill in a factory in Beloit, Wisconsin.

Both Halladay and Wheeler were awarded gold medals at the 1876 Philadelphia Centennial Exposition in recognition of their inventions. However, many other manufacturers were offering windmills and neither of the pioneers was able to

Figure A-5. European post mill.

dominate the field. In 1889, there were an estimated 77 windmill factories employing over eleven hundred workers, with sales of over four million dollars in the United States. Almost a thousand patents had been issued. As much as 25 percent of nontransportation energy may have been provided by windpower. The three big markets were the railroads, irrigation and stock watering, and pumping of water for well-to-do families who wanted running water for their bathrooms.

The "American" windpumpers had many variations and features, but most were small (less than 1 hp) in capacity, and of the multi-blade type. Though inefficient as pumps, those machines were both cheap and reliable. They were less costly to produce and easier to erect than the earlier "European" windmills and required less maintenance and attention.

Figure A-6. Artist's sketch of tilting at a post mill.

The culmination of the Nineteenth Century windpower revolution was the dramatic installation of dozens of windpumpers at the World's Columbian Exposition in Chicago in 1893. The windpumpers competed successfully for attention with Pullman's sleeping cars, Edison's phonograph, and other mechanical and electrical innovations. They not only pumped water, but also shucked corn, ran lathes and sewing machines, and showed how the wind could lighten other familiar chores.

Most of the machines for the Exposition had been built in factories in the small towns of the Chicago area. In many cases, the windpumper companies had the only factories in those rural areas and were the largest employers. Workmen were proud of their craftsmanship and followed their products to Chicago to cheer. Some of the more popular machines were immortalized by photographs and paintings, as well as popularized on shaving mugs and household items—certainly a high point in the historic evolution of windpower.

From 1854 until now (units similar to the early machines are still being built and reconditioned today) there have been approximately six million small windpumpers built in the United States. An estimated 150 thousand are still in service. The established windpumper companies which still exist are Aeromotor, Dempster, and Heller-Aller.

Electricity generating wind turbines evolved primarily in the Twentieth Century although there was considerable experimentation during the second half of the

Figure A-7. Artist's version of tower windmill.

Nineteenth. The Smithsonian Institution displays a small wind conversion device which produced enough electricity in the 1860s to power an incandescent light-bulb. That invention earned a patent for Moses G. Farmer.

Nothing much came of Farmer's invention, but the Danes were more persistent. The Danish government started developing large wind systems in the 1890s and as much as 25 percent of that country's industrial power (mechanical and electrical) came from approximately 25 hundred windmills by the turn of the century. Professor Poul la Cour, a teacher at the high school in the town of Askov, is credited with pioneering work on electricity generating WECS. He began his aerodynamics experiments in 1891, and by 1908 more than 70 machines (with electricity generating capacities of 5-25 kW) had been installed. Prof. la Cour initially stored his electricity by electrolyzing sodium hydroxide and using the

Figure A-8. Generic Dutch windmill.

gases for lighting in the high school. Later, batteries (or "lead accumulators") were utilized. As a result of that work, the Danish Wind Electricity Company was formed in 1903 and by 1918 there were about 120 of the 75 ft diameter, four-blade wind turbines in operation to help keep Denmark going during World War I when their imported diesel fuel was shut off.

After World War I, serious wind-electric experimentation started in the United States. Although electric lines weren't yet available to most rural homes and farms, batteries were, and people were beginning to enjoy the benefits of electricity.

Much of the early wind-electric work was by do-it-yourselfers interested in recharging their DC batteries. To generate electricity, improvements in the traditional windpumpers had to be made to enable them to run much faster.

Figure A-9. Typical European cluster of windmills.

That led to use of airplane-type propellers and to specially designed factory-built machines. By the 1920s, several "wind turbine generators" were available and efforts were being made to develop more efficient, higher speed rotors with fewer blades made from thinner metal and wood.

The first commercially available machines were the Aeroelectric "farm lighting plants" rated at 1 kW, which were produced by an old line (founded in 1860) windpumper company in South Bend, Indiana—the Perkins Corporation.

In 1922, M.L. and J.H. Jacobs started designing wind-electric systems in Montana. After building and testing units for several years on their own and neighboring ranches, Jacobs Wind Electric Company established a plant in Minneapolis in 1930. During the next thirty years, Jacobs built "tens of thousands" of 2 kW and 3 kW three-bladed propeller type DC "battery storage wind electric systems." (See Figure A-12.) They shut down in 1960 when extension of power lines by REA reduced demand. Marcellus Jacobs started his comeback in 1980, at age 77, by offering a 10 kW, three-blade HAWT for grid-connected AC electricity. His new company quickly reestablished itself as a leader in small WECS. His new entry was welcomed by WECS dealers and customers because of the record of successful operation of his smaller units, many of which are still performing.

Figure A-10. Larger Dutch-type windmills.

John and Gerhard Albers got started in Cherokee, Iowa, at about the same time as the Jacobs brothers were experimenting in Montana. They opened a plant in Sioux City, Iowa to produce the 200 Watt Wincharger for 6, 12, 32, and 110 volt DC battery charging. The Albers Company was acquired by the Zenith Radio Company and later became the Winco Division of Dyna Technology, Inc. The 12 volt Wincharger is still being sold today, the only pioneer still around.

In addition to Perkins, Jacobs, and Wincharger, other familiar world-class small wind-electric companies included Aerowatt (France), Dunlite (Australia), and Elektro (Switzerland). Today, in addition to the new Jacobs, such com-

Figure A-11. Typical American windpumper.

panies as Aeropower (California), Bergey Wind Power (Oklahoma), Enertech (Vermont), North Wind Power (Vermont), and Windworks (Wisconsin) are evolving as leaders in the very small (under 20 kW) segment of the wind industry.

Modern electricity generating wind energy conversion systems have benefited from the long experience with windmills, windpumpers, battery chargers, and very small wind-electric machines. However, in scaling up to WECS of substance, direct transfer of that technology has proved to be difficult, time consuming, expensive and fraught with failure and frustration.

In addition to the higher speed two- and three-blade propeller developments in the United States in the '20s, vertical axis inventions by Georges J.M. Darrieus

Figure A-12. Early Jacobs Wind Electric System.

(France), Anton Flettner (Germany), and Captain Sigurd N. Savonius (Finland) in that same decade led to some of today's more important technology.

Darrieus conceived several vertical axis designs (straight blades, curved "troposkein"-shaped blades and other combinations) in 1925, and these innovations can be traced through European, Canadian, and U.S. evolution to today's commercially available Darrieus type VAWTs and several advanced national WECS research programs.

Darrieus also worked on more conventional propeller type HAWTs, and in 1929 the Campagnie Electro-Mecanique erected a two-blade, 66 ft diameter research machine at Bourget, France.

Flettner attempted to harness the thrust exerted by a cylinder spinning in a windstream (the Magnus effect) and crossed the Atlantic in a rotor ship (Badden-Badden) in 1925. (See Figure A-13.) In 1926, he built a 66 ft diameter four-armed "windwheel" which resembled propeller type HAWTs, except that each of the arms was made up of three 16 ft long spinning aluminum shells. The prototype was rated approximately 30 kW at 23 mph. A simpler version of the Flettner concept was prototyped by Public Service Corporation of New Jersey in 1933 at Burlington, New Jersey. That single cylinder was 90 ft high and 18 ft in diameter. The ultimate goal was to have twenty flat cars with those cylinders running around a 3000 ft circular railroad track while generating electricity. The prototype proved inefficient and the scheme was not tried.

Captain Savonius developed a vertical cylinder sliced in half from top to bottom, the two halves being pulled apart by about twenty percent of the diameter, much like a cup anemometer. (See Figure A-14.) Recirculation of flow helped rotate the cylinder by pushing the half of the cylinder which was moving backwards upwind. The Savonius rotor worked, and in fact is still widely used in experimental work, but used too much material to be cost-effective.

The first known large WECS was built at Balaclava, near Yalta, on the Black Sea, in 1931. (See Figure A-15.) The horizontal axis rotor had a diameter of 100 ft and was automatically controlled by pitching the blades and moving the rotor to face the wind. The entire structure was rotated by moving a stiff tail strut on a carriage mounted on a circular track on the ground. The Russian machine was rated 100 kW at 24.6 mph. Its induction generator was grid connected to a 20 MW steam generation station some 20 miles away at Savastopol. The machine reportedly worked well for ten years and led to an unsuccessful proposal to build a much larger system rated at 5 MW. However, the Russians have reportedly proceeded with construction of many intermediate scale HAWTs. (See Figure A-16.)

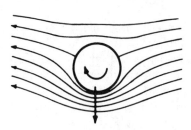

Figure A-13. Flettner's Magnus effect.

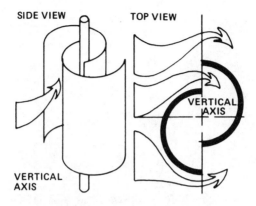

Figure A-14.   The Savonius rotor concept.

Figure A-15.   Early Russian 100 kW HAWT prototype.

Figure A-16. A variety of Russian windmills.

Shortly after the Russian machine was erected, Professor Hermann Honnef proposed in Berlin in 1932 to build a gigantic multi-rotor structure rated at 50 MW, but his proposal was repeatedly rejected by German officials (and later by Marshall Plan administrators) because wind was not thought to be a practical source of power. (See Figure A-17.)

At about the same time the Russian machine was being dismantled at Yalta, the famous Smith-Putnam machine (the first ever megawatt scale WECS) was being turned on at Grandpa's Knob in the windy Green Mountains near Rutland,

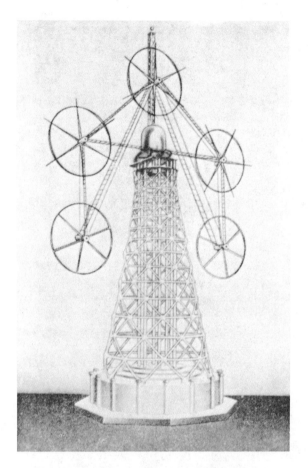

Figure A-17. Honnef's proposed 50 MW multi-rotor.

Vermont. (See Figure A-18.) That 1.25 MW capacity (rated at 30 mph) HAWT culminated work begun (partially inspired by the Russian adventure) in 1934 by Palmer C. Putnam. The downwind machine had two 87.5 ft long stainless steel blades which rotated at 28.7 rpm to drive a synchronous generator connected to the Central Vermont Public Service Company system. The first turn was on August 29, 1941, and electricity was first fed into the Vermont system on October 19. A bearing failed in 1943 and, because of the war, the machine was out of service until early 1945. Then, after about 1100 hours of operation and under close scrutiny by the technical team, a blade was lost in March of 1945. Although all concerned wanted to rebuild the rotor, the war was over and oil was once again available and cheap, and the project was abandoned. The pioneering project

Figure A-18. The Smith-Putnam 1.25 MW HAWT at Grandpa's Knob.

was considered a technical success and, fittingly, on July 29, 1981 Palmer Putnam, Beauchamp E. Smith—president of S. Morgan Smith Co. of York, Pa., which funded the prototype project and brought together other important participants such as General Electric (electrics), Budd Manufacturing Co. (blades), Wellman Engineering Co. (drive train), and American Bridge Company (tower)-and seven members of the project team (John B. Wilbur, chief engineer; Carl J. Wilcox, project engineer; Grant T. Voaden, chief test engineer; Stanton D. Dornbirer, chief of field assembly; Homer J. Steward, aerodynamicist; Myle J. Holley, test engineer; and Hurd C. Willett, meteorologist) were honored at the 40th anniversary of their historic achievement during a DOE/NASA workshop on large horizontal axis wind turbines held in Cleveland.

The Grandpa's Knob adventure (which is described in Palmer Putnam's 1948 book, *Power From The Wind*, published by Van Nostrand Reinhold Company, New York) was a major milestone in efforts to develop large-scale utility-integrated WECS which are still underway today.

An immediate result of the Smith-Putnam experiment was a ten year effort by Percy H. Thomas, an engineer with the U.S. Federal Power Commission, to design two multi-rotor machines, one 6.5 MW and one 7.5 MW in capacity, and to try to convince congress to fund a prototype to demonstrate economic feasibility of large WECS in the 5–10 MW range. (See Figure A-19.) By 1951, there was some interest in funding a Federal Power Commission prototype, but the Korean War brought the project to a stop.

Meanwhile, the Danes were active in the '40s with 60 kW and 70 kW capacity HAWTs built by the F.L. Smidth Company. Between 1941 and the early '50s, 12 of the 60 kW units (rated at 22 mph with a two-blade, 57 ft diameter rotor) were built. Between 1943 and the '50s, six of the 70 kW units (rated at 19 mph with a three-blade, 79 ft diameter rotor) were constructed. A unique feature of both Smidth models was that they ran at variable speed to best utilize the wind.

The Danes proceeded, in 1953, to build a modified version of the earlier 60 kW and 70 kW Smidth designs (this one a 43 ft diameter HAWT rated 45 kW) which was tested through the end of the decade. However, the major Danish accomplishment was installation in 1958 of the famous 200 kW capacity (rated at 36 mph) HAWT near Gedser on the Island of Falster in southernmost Denmark in the Baltic Sea. (See Figure A-20.) That prototype was the culmination of a major effort initiated in 1947 by senior engineer Johannes Juul of the private electric company SEAS, and carried out by the wind power committee of the Danish Association of Electricity Supply Undertakings. The three-blade upwind rotor swept an area with a diameter of 79 ft at 30 rpm to drive an induction generator which, in turn, was interconnected to the utility's electrical system. The machine operated until 1967, when it was shut down because of lack of interest caused by a belief that wind generated electricity was about twice as expensive as fossil fueled generation. However, because of its historic technical significance, and with a newfound interest in wind caused by the oil embargo, the Gedser HAWT was refurbished and put back into service in the fall of 1977 as a joint project of the Danish and U.S. governments. The wind turbine was monitored and tested by Riso National Laboratory, Danish Ship Research Laboratory, and Technical University of Denmark until 1979 when U.S. funding expired.

In 1954, an innovative horizontal axis wind turbine (rated 100 kW at 33 mph) was built at St. Albans, England. The 80 ft diameter rotor had two hollow blades which operated at 95 rpm and pumped air which was utilized instead of gears to

Figure A-19. The Percy Thomas multi-rotor concept.

Figure A-20. The famous Danish Gedser 200 kW HAWT.

transmit the propeller power to the generator. The machine was built by Great Britain's Enfield Cable Company and was designed by French engineer Andreau (it is sometimes referred to as a French innovation). The Enfield-Andreau air pump scheme proved inefficient compared to mechanical couplings and the HAWT prototype was eventually moved to Algeria. (See Figure A-21.)

In 1955, John Brown Company built and installed an experimental three-bladed HAWT (rated 100 kW at 35 mph) on Cape Costa in the Orkney Islands for the North Scotland Hydroelectric Board. (See Figure A-22.) The rotor was 50 ft in diameter and operated at 130 rpm. However, that English prototype,

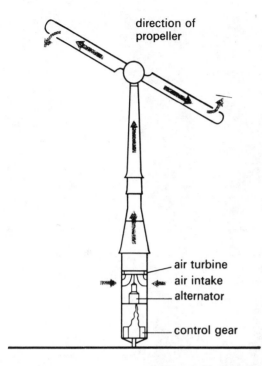

Figure A-21. Enfield-Andreau ducted rotor.

coupled to a diesel powered electric utility grid, operated for less than a year because of technical problems.

The Germans, led by Professor Ulrich Hutter, performed some of the most important large-scale system work in the '50s. An impressive HAWT prototype ran from September, 1957 until August, 1968. That machine (rated 100 kW in only 18 mph winds) used lightweight 55 ft composite glass fibre and carbon fibre reinforced plastic blades which could have their pitch changed at higher wind speeds to keep the propeller rotation constant. (See Figure A-23.) Another innovation was the use of a simple hollow pipe tower supported by guy wires. When the machine was dismantled, the blades were found to be sound and undamaged after ten years of operation.

The French were also active in the '50s. The Bureau d'Etudes Scientifiques et Techniques of France engineered an 800 kW (rated at 36–37 mph) HAWT for the French Electricity Authority and a prototype was erected southwest of Paris near Nogent Le Roi in 1958. (See Figure A-24.) That three-blade machine had a rotor diameter of 105 ft and operated at 47 rpm. The rotor and nacelle were fixed to a tower which, in turn, was mounted on a turntable on the top of a

Figure A-22. The John Brown 100 kW prototype.

second Eiffel-like tower anchored to the ground. The entire upper structure was yawed to track the wind direction. A blade broke off and destroyed the machine in the 1960s and it was not rebuilt. Two smaller, three-blade machines were reportedly built in southern France near St. Remy des Landes—one with a 70 ft rotor diameter operating at 56 rpm to provide 130 kW at 28 mph, and a second with a 100 ft diameter rotor which generated 300 kW at 37 mph.

A major wind energy accomplishment was the reinvention in 1966 of Darrieus type vertical axis wind turbines at the National Research Council of Canada's Low Speed Aerodynamics Laboratory near Ottawa. Raji Rangi and Peter South, with support of laboratory manager Jack Templin, initiated work which led to major VAWT development and commercialization efforts in Canada, the United States, Europe, Australia, and New Zealand. By 1974, the NRC had involved private Canadian companies—Bristol Aerospace in Winnipeg and Dominion Aluminum Fabricating (DAF-Indal) in Toronto—in their small (1–4 kW) VAWT program.

Figure A-23. German Hutter–Allgaier 100 kW HAWT.

OIL EMBARGO RESPONSE—THE 1970s

In response to the first oil embargo and its resultant dramatic price increases, the United States and many of the world's industrialized nations initiated substantial taxpayer-funded programs to "research, develop, demonstrate and encourage commercialization of wind energy conversion systems."

Australia, Austria, Canada, Denmark, England, Germany, Holland, Ireland, Japan, New Zealand, Spain, Sweden, and the United States had active programs. Multilateral agreements provided for exchange of information and cooperative developments. In most countries, the emphasis was on megawatt scale electricity generating machines.

The United States Federal Wind Energy Program was initiated in 1973 under the auspices of the National Science Foundation (NSF) and picked up by the Energy Research and Development Administration (ERDA) in 1975 and its successor organization, the Department of Energy (DOE), when it was established

Figure A-24. The French 800 kW HAWT near Paris.

in 1977. The U.S. Departments of Agriculture (USDA) and Interior (DOI) also initiated important wind energy programs.

The NSF/ERDA/DOE programs were executed through contractors responsible for particular elements. The National Aeronautics and Space Administration (NASA) headed the R&D program on large (over 100 kW) horizontal axis wind turbines through its NASA Lewis Research Center in Cleveland, Ohio. Sandia National Laboratories in Albuquerque, New Mexico, managed the government efforts in Darrieus-type vertical axis wind turbines. DOE's Rocky Flats Plant near Golden, Colorado administered work on small (less than 100 kW) wind energy conversion systems directed toward dispersed farm and home applications. DOE's Pacific Northwest Laboratory in Richland, Washington took on the task of locating, understanding and explaining the nation's wind resources.

The DOE/NASA effort was toward bigger and lower cost HAWTs. Their program is shown in Table A-1.

Table A-1. DOE/NASA HAWT Research Program.

NASA R&D		PROTOTYPE		
HAWT	CAPACITY	BUILT BY	INSTALLED AT	FIRST TURNED
MOD-0	100 kW @ 14.5 mph	NASA	Sandusky, OH	Sept., 1975
MOD-0A (four machines)	200 kW @ 18.3 mph	Westinghouse	Clayton, NM	March, 1978
		Westinghouse	Culebra, PR	Jan., 1979
		Westinghouse	Block Isl., RI	Oct., 1979
		Westinghouse	Oahu, HI	May, 1980
MOD-1	2.0 MW @ 25.7 mph	G.E.	Boone, NC	Sept., 1979
MOD-2 (four machines)	2.5 MW @ 19.5 mph	Boeing	Goldendale, WA	Dec., 1980
		Boeing	Goldendale, WA	Feb., 1981
		Boeing	Goldendale, WA	May, 1981
		Boeing	Medicine Bow, WY	early 1982
MOD-5A	6.2 MW	G.E.	design study only	
MOD-5B	7.2 MW	Boeing	design study only	

MOD-0 (see Figure A-25) has two downwind, 62.5 ft long blades. The test bed machine was installed at the NASA Lewis Research Center in 1975 to prove the concept and to serve as a prototype test station for new developments. It has a synchronous generator which is interconnected with the local Ohio Edison Company utility grid. Late in 1981, the original rigid truss tower was changed to a softer tubular tower for continuing research.

MOD-0A (see Figure A-26) has two downwind blades of the same size as MOD-0. The rotor operates at 40 rpm to drive a 250 kVA synchronous AC three-phase generator. The first prototype is interconnected to the municipal utility grid in Clayton, New Mexico. The second unit is on an island off the mainland of Puerto Rico and is monitored by remote control by the Puerto Rico Electric Power Authority. The third unit is connected to the local Block Island Power Company which has no connection to the mainland and is totally dependent on oil. The fourth, and most successful of the MOD-0As, is at Kahuku Point where it is interconnected to the Hawaiian Electric Company's grid. Salt spray corrosion caused replacement of the Hawaii prototype's blades in mid-1982.

MOD-1 (see Figure A-27) has two downwind blades which define a swept rotor diameter of 200 ft intended to operate at a constant 35 rpm. The R&D prototype machine was expected to serve the Blue Ridge Electric Membership Corporation (a cooperative) and save on their energy purchases from Duke Power Company. In 1980 the rotor speed was reduced to 23 rpm (1.5 MW) to reduce

Figure A-25. The U.S. DOE/NASA 100 kW MOD-0 test bed.

noise problems. However, many problems have kept this NASA prototype from performing, and the test program is in the process of being terminated.

MOD-2 (see Figures A-28 and A-29) was a cooperative effort between DOE/ NASA and the Bonneville Power Administration and, later, the Department of Interior. In addition to testing Boeing-built HAWTs, the three unit cluster in the Goodnoe Hills area near Goldendale, Washington in the Columbia River Gorge was intended to demonstrate and test many aspects of windfarms. The fourth

Figure A-26a. The DOE/NASA 200 kW MOD-0A, Clayton, NM.

Figure A-26b. POE/NASA 200 kW MOD-0A, Block Island, RI.

unit, purchased by DOI's Bureau of Reclamation for their Medicine Bow, Wyoming project, was also intended to be part of a mini-windfarm. The MOD-2 has two 150 ft long upwind blades which operate at 17.5 rpm to drive a 3125 kVA synchronous AC generator. However, serious mechanical damage to one of the HAWTs, caused by a runaway condition of its rotor in June, 1981, just eleven days after dedication of the 7.5 MW cluster, stopped all testing until the cause could be determined and all three machines retrofitted. The two undamaged machines were later returned to service for additional testing.

Figure A-26c. DOE/NASA 200 kW MOD-0A, Culebra Island, PR.

In addition to the four government purchased MOD-2s, Pacific Gas and Electric Company (San Francisco) bought one as part of their wind research efforts. Their prototype is in Solano County, 40 miles northeast of San Francisco.

The two *MOD-5s* were considered by DOE and NASA to be the ultimate in HAWT technology and to have a chance to be cost competitive with conventional generators if produced in volume. The General Electric MOD-5A design (6.2 MW) envisioned a two-blade, upwind rotor with a 400 ft diameter swept area. The Boeing MOD-5B design (7.2 MW) has a 420 ft diameter rotor, also with two blades located upwind of the nacelle. However, with U.S. DOE budget cuts, it is doubtful that either will be prototyped.

Figure A-26d. DOE/NASA MOD-0A, Oahu, HI.

The DOE/Sandia Laboratories VAWT research program started with a hand-built, 15 ft diameter Darrieus prototype, with Savonius starters, in May, 1974 to demonstrate the concept. (See Figure A-30.) Later, continuous aluminum blades replaced multi-segment blades and significantly improved efficiency and costs. (See Figure A-31.)

Sandia then built a larger prototype at their Albuquerque test site which utilized Kaman Aerospace composite blades. (See Figure A-32.) The machine had a rotor 55 ft high and 55 ft in diameter and was tested with both two (50 kW) and three (60 kW) strut-reinforced blades. Later, Alcoa supplied larger (24 in. vs. 21 in.) chord extruded aluminum blades which eliminated the need for reinforcing struts and allowed the prototype to be much more aerodynamically efficient (45 percent vs. 38 percent) and raise peak power (from 50 kW to 60 kW with two blades), at lower cost. (See Figure A-33.)

Figure A-27. DOE/NASA 2 MW capacity MOD-1 near Boone, NC.

Alcoa Laboratories won a DOE contract in 1977 to design a second-generation, lower cost version of the larger (55 ft diameter, or 17 metres) Sandia prototype which became known as "the low cost 17 metre VAWT." (See Figure A-34.) That contract culminated in successful prototype installations at Rocky Flats, Colorado (August, 1980), Bushland, Texas (December, 1980), and Martha's Vineyard, Massachusetts (February, 1981). The "low cost" version raised the rotor height from 55 ft to 83 ft and resulted in a power capacity of over 100 kW at 30 mph.

Figure A-28. DOE/NASA 2.5 MW MOD-2 at BuRec's Medicine Bow Site.

The encouraging "low cost 17 metre VAWT" prototypes and the Sandia technology dissemination program brought forth several companies attempting to commercialize their version of the technology. By early 1982, Alcoa, Flow Industries, and Forecast Industries had announced plans. Commercial versions of the smaller Sandia prototype were offered earlier by Dynergy and Tumac.

Before the DOE budget cuts, Sandia had planned larger VAWTs and was negotiating a contract for a nominal megawatt scale development program (MOD-6V) with Alcoa Laboratories. With the new reality, Sandia is concentrating their efforts on testing their five prototypes, refining their technology base, and assisting industry in privately funded R&D efforts.

Figure A-29. The PG&E MOD-2 in California.

DOE's Rocky Flats R&D program funded design and prototype small WECS of 1 kW, 4 kW, 8 kW, 15 kW, and 40 kW capacity, all rated at 20 mph. The 15 kW and 40 kW contracts led to technology currently being commercialized by Enertech (20 kW version of the 15 kW HAWT) and available to others. In addition, United Technologies Research Center (UTRC) 15 kW HAWT and Kaman Aerospace (HAWT) and McDonnell Douglas (straight-blade VAWT) 40 kW research machines are being tested at the Rocky Flats Test Center. The first DOE/Sandia "low cost 17 metre VAWT" (100 kW) is also being tested at Rocky Flats. (See Figures A-35–A-37.)

In Canada, the National Research Council's (NRC) Low Speed Aerodynamics Laboratory had been working on small VAWTs since 1966. After the 1973 oil

Figure A-30. Original Sandia 5 metre VAWT prototype.

embargo, emphasis was reoriented toward larger systems capable of supplying electricity to remote communities and to provincial utility networks.

Units up to 500 kW in capacity have been commercialized by Canadian firms (Bristol Aerospace of Winnepeg and DAF-Indal of Toronto) utilizing the technology and experience from building prototypes for NRC. Several 50 kW units, built by DAF-Indal, were installed in Newfoundland, British Columbia, Saskatchewan, and Manitoba in Canada and in both Australia and the United States. (See Figures A-38 and A-39.) Their most impressive 50 kW accomplishment was a 1980 installation on a peninsula in front of the Romero Overlook Visitors Center at San Luis Reservoir for the California Department of Water Resources and Pacific Gas and Electric Company. A similar VAWT (50 kW, 55 ft × 37 ft rotor) was later fabricated by DAF for installation at San Gorgonio Pass by Southern California Edison Co.

A larger (120 ft high × 80 ft diameter, two-blade rotor at 38 rpm) Darrieus type VAWT was built by DAF-Indal for NRC and Hydro Quebec and installed in

Figure A-31. Retrofitted Sandia 5 metre VAWT.

the summer of 1977 on one of the Magdalen Islands in the Gulf of St. Lawrence. (See Figure A-40.) That prototype, which was rated 230 kW in 30 mph wind, spun out of control and was destroyed in July, 1978 when maintenance work-men left the machine unattended without brakes or electrical load in the belief that VAWTs are not self-starting. That VAWT proved to be self-starting, and there was no way to stop it when its spoilers failed to deploy. The prototype was rebuilt and reinstalled in September, 1979 and began electrical generation again in March, 1980. It has worked well since that time with maximum power exceeding predictions by a significant margin. DAF is now attempting to com-mercialize second-generation versions of that prototype.

The Canadian R&D effort has moved toward multi-megawatt size VAWTs and is known as project AEOLUS. NRC has contracted with a consortium of Canadian companies (including Canadaire and Shawinigen Engineering) to design and build

Figure A-32. Original Sandia 17 metre VAWT prototype.

a much larger VAWT in cooperation with Hydro Quebec. That two-blade machine is expected to be rated at approximately 3.4 MW.

In Denmark, the Ministry of Commerce and the Danish electric utilties funded their wind program in 1977 with major emphasis on development of two intermediate size HAWTs on the windy seacoast near Nibe. In addition, the U.S. Department of Energy partially funded restoring and testing the famous 200 kW upwind HAWT which operated successfully from 1957 to 1968 at the town of Gedser. The Gedser mill was refurbished and put back into service in November, 1977 and tested until April, 1979.

The first of the two Nibe HAWTs was put into service in August, 1979 and the second was commissioned in March, 1980. The machines at Nibe are identical in size but the first (Unit A) has its rotor blades supported by stays and regulated with four discrete pitch angles. The second (Unit B) allows the blades to be self-supporting and provides full pitch control for power regulation. The

Figure A-33. Retrofitted Sandia 17 metre VAWT.

three-blade, 131 diameter rotors operate at 33 rpm to drive asynchronous induction generators which are rated 630 kW at 29 mph. (See Figure A-41.)

Based on its experience as the reinforced plastic blade supplier for the Nibe machines, Volund A/S, one of Denmark's leading manufacturing and exporting companies, has developed a three-blade, downwind, variable pitch HAWT rated at 265 kW in 29 mph winds. The machine has a 93 ft diameter rotor and operates at 28 rpm in low winds to drive a 58 kW induction generator and at 42 rpm in higher winds to drive the 265 kW induction generator. The innovative two generator scheme is intended to provide higher electric efficiency. While proceeding with commercialization of the 265 kW model, Volund is designing two-speed 530 kW (125 ft diameter) and 750 kW (150 ft diameter) models.

Figure A-34. Alcoa/Sandia low cost 17 metre VAWT on Martha's Vineyard.

Independent of government programs, an adventurous Danish "university" consisting of the three Tvind Schools of Denmark (a seminary, a commuter high school, and a junior college) built the world's largest (at that time) WECS. Work by students and faculty started in May, 1975 and was completed in January, 1978. The three-blade rotor sweeps an area with a diameter of 177 ft and drives a 2200 kVA three-phase synchronous AC generator at variable speeds to maximize efficiency. The do-it-yourself downwind HAWT, although containing a 2 MW generator, is restricted to 400 kW of 50 Hz power to the local network (through an inverter), and 500 kW of variable frequency AC for heating the school. It is working well in a nonautomatic mode at Ulfborg on the windy west coast of Jutland, Denmark. (See Figure A-42.)

Figure A-35. Kaman Aerospace 40 kW HAWT at Rocky Flats.

The Federal Republic of Germany wind energy R&D program was initiated in the summer of 1976 and is known as GROWIAN (the acronym for Grosse Windenergie-Anlage).

Their first prototype was a small 18 ft diameter Darrieus type VAWT, with Savonius buckets for starting, as part of their SWECS program for special uses. The major efforts, however, have been with large HAWTs with progress scheduled as shown in Table A-2.

The Voith-Hutter machine has a 160 ft diameter, two-blade rotor which operates at 37.1 rpm and is based on Professor Ulrich Hutter's earlier 100 kW (rated at 18 mph) advanced HAWT prototype which operated successfully from 1957 to 1968 near Stuttgart. It was installed at Stotten in southern Germany. Voith Transmissions is an established German firm which produces water turbines and other power plant equipment. The composite reinforced plastic blades were fabricated by Allgaier Company which had pioneered composite blades in the '50s and demonstrated them on the first Hutter prototype.

Messerschmitt-Boelkow-Blohm (MBB) built a one-blade HAWT demonstration prototype which is one-third the size of the ultimate 5 MW HAWT. Its single counterbalanced blade rotates to define a swept area with a 158 ft diameter and drive a 350 kW generator. The full scale unit—GROWIAN II—has a single blade to provide a rotor diameter of 476 ft. Its counterbalance extends 66 ft

Table A-2. West German HAWT R&D Schedule.

| GERMAN R&D | | PROTOTYPE | |
HAWT	CAPACITY	BUILT BY	TARGET INSTALLATION
Voith-Hutter	265 kW @ 20 mph	Voith Transmission	1981
MBB Demo	350 kW @ 25 mph	Messerschmitt-Boelkow-Blohm	1981
GROWIAN I	3 MW @ 26.5 mph	Maschienenfabrik Augsburg-Nurmberg	1982
GROWIAN II	5 MW @ 25 mph	Messerschmitt-Boelkow-Blohm	1986

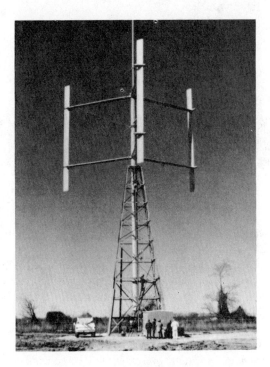

Figure A-36. McDonnell Douglas 40 kW VAWT at Rocky Flats.

Figure A-37. DOE/Sandia low cost 17 metre VAWT at Rocky Flats.

from the hub which is at a height of 394 ft. From the ground to the tip of the blade, when it is straight up, is 632 ft.

The more conventional two-blade 3 MW GROWIAN I was built on the windy coast of Schleswig-Holstein, and feeds its electricity into the 20 kV network which serves Hamburg. The contractor to the West German Ministry for Research and Technology is Maschinenfabrik Augsburg Nurmberg (MAN). The 329 ft diameter rotor operates at 18.5 rpm and drives an asynchronous AC generator at variable speeds.

In the Netherlands, the Dutch government initiated a wind energy program in February, 1976. The first prototype was a small 15 ft diameter Darrieus type

Figure A-38. Early DAF-Indal Canadian 50 kW VAWT.

VAWT designed and built by Fokker Aircraft which has been tested since 1977. However, the major effort has been with large HAWTs.

The Netherlands Energy Research Center contracted with FDO, a wholly owned subsidiary of VMF-Stork N.V., to develop and construct an intermediate size HAWT. The prototype, which was installed in 1978 at Petten, has an 82 ft diameter two-bladed upwind rotor with variable pitch which drives a 300 kW DC generator. An inverter then turns the DC into AC. In addition to the prototype (designated 25 HAT MK1) FDO has designed three additional HAWTs. The four designs are shown in Table A-3.

Table A-3. Dutch HAWTs by FDO

| | | ROTOR | |
FDO MODEL	ELECTRICAL GENERATOR	BLADES	DIAMETER
25 HAT MK1	300 kW DC	2	82 ft
25 HAT MK2	300 kW AC	3	82 ft
25 HAT MK3	300 kW AC	3 (fixed)	82 ft
50 HAT MK1	1 MW AC	2	164 ft

Figure A-39. DAF-Indal VAWT in U.S. test program.

The National Swedish Board for Energy Source Development funded a prototype two-blade downwind HAWT which was built by Saab-Scania Aerospace and put into service on April 28, 1977. That 65 kW machine, rated at 22 mph, has a rotor diameter of 59 ft and operates at 77 rpm. (See Figure A-43.) It is located at Kalkugnen by Alvkarleby close to the Baltic Sea. The Swedish State Power Board was operating and monitoring the prototype until early 1981, when it was damaged by a crane during repair operations. Saab-Scania has conducted analytical studies of 1 MW (at 28 mph) and 4 MW (at 28 mph) HAWTs with 112 ft and

Figure A-40. DAF-Indal 230 kW VAWT on Magdalen Islands.

225 ft diameter rotors, respectively. The 1 MW rotor would operate at 70 rpm and the 4 MW at 21 rpm.

A 2 MW capacity (at 28 mph) two-blade upwind HAWT is being built on the southwestern coast of the Island of Gotland at Nasudden. Karlstads Mekaniska Werkstad, in cooperation with Erno, an aerospace subsidiary of the German VFW-Fokker group, is the contractor for the machine for the State Power Board.

Figure A-41. One of the twin 630 kW Danish HAWTs at Nibe.

Figure A-42. The Danish 2 MW Tvind prototype.

Figure A-43. The Swedish Saab-Scania 65 kW HAWT prototype.

The machine has a rotor diameter of 254 ft and will operate at 25 rpm to drive an induction generator.

Separately, the government proceeded with a 3 MW at 31 mph (WTS-3) HAWT built by a consortium which includes Sweden's shipbuilding giant, Karlskronavarvet, as the prime contractor, and Germany's Thyssen Henschel, and the United States' Hamilton Standard Division of United Technologies. That two-blade downwind machine was built in 1981 and 1982 at Maglarp on Sweden's southern coast. Its 261 ft diameter rotor drives a synchronous AC generator.

With Hamilton Standard as the prime contractor, a similar venture has led to the construction of a 4 MW capacity (WTS-4) HAWT as a "system verification unit" for the U.S. Department of Interior's Bureau of Reclamation wind project near Medicine Bow, Wyoming. The WTS-4 is sited near the NASA MOD-2 proto- type. The project is the start of what BuRec hopes will be a 100 MW wind farm integrated with their hydroelectric facilities.

The United Kingdom's Central Electricity Generating Board (CEGB) has been actively pursuing both horizontal and vertical axis machines, resource assessment, and prospects for off-shore siting.

The English VAWT work has been mainly with a straight-blade scheme devel- oped by P.J. Musgrove of Reading University. Small scale designs have been suc- cessfully prototyped and current government funded work is underway by an industrial consortium headed by Sir Robert McAlpine and Sons Limited to develop 325 ft diameter machines rated at 4.4 MW. Design of an 82 ft diameter, 130 kW test-bed machine (quarter scale of the planned big one) was completed and prototyped.

A privately financed three-blade HAWT with a 56 ft diameter rotor was built in 1977 at Aldborough Manor in North Yorkshire by Sir Henry Lawson-Tancred Sons and Company. (See Figure A-44.) That innovative unit has a hydraulic drive instead of gears and was intended to serve as a prototype for a 1.5 MW capacity HAWT four times as large.

Two government sponsored HAWTs are being pursued for installation on the windy Orkney Islands for evaluation by the North of Scotland Hydroelectric Board. The site reportedly has winds which average 26 mph. The government contractor is Wind Energy Group, a consortium of Taylor Woodrow Construc- tion, British Aerospace Dynamics Group, and GEC Power Engineering. The first HAWT is expected to have two 33 ft long blades and be rated at 250 kW. The second, scheduled for construction in 1983, will have a rotor diameter of 196 ft and be rated at 3 MW.

TODAY'S WECS EVOLUTION—THE 1980s

At the time of this writing (April, 1982), the status of the WECS industry is still evolving. No industry leader has yet emerged. No WECS has yet proven itself in service for more than a few months. Contraction of the U.S. Department of Energy's wind program has caused the professional "government contractors" to fall by the wayside. The recession has caused other companies to put their WECS programs on hold or to slow them considerably. At the same time, dramatic failures of both privately funded and government sponsored research prototypes has led to a sobering reevaluation of the state-of-the-art and a new respect for the complexities and safety issues inherent with high speed rotating equipment.

Figure A-44. English 30 kW prototype at Aldborough.

The glamor and excitement has been left behind, and the remaining participants are settling in for the hard work of proving and refining their hardware and riding out the recession while getting their systems ready for the market explosion expected in the mid-'80s.

A few U.S. companies are producing WECS with generating capacities in excess of 20 kW (see Chapter 2 and the section on the producers) but the industry is still searching for its identity and its proper role in the nation's energy mix. The opportunity is still alive for innovation and for cost-effective performance-reliable, and safe wind energy conversion systems of almost all sizes and types. Future history will record if there are any winners waiting in the wings.

Appendix B
Acronyms and Definitions

AC. Alternating current.

Airfoil. A curved surface designed to create lift as air flows over its surface.

AkWh. Annual kilowatt-hours.

Alternating Current (AC). Electric current which alternates direction of electricity flow, used in utility networks for most residential, commercial, and industrial requirements.

Ampere (amp). A measure of the strength of electric current flowing in a circuit.

Ampere-Hour (amp-hour). Measure of energy potential. One amp-hour will provide one amp of current for one hour at the given voltage. Voltage X amperage gives power in watts and amp-hours X voltage = watt-hours of energy.

Anemometer. An instrument for measuring windspeed (velocity).

AOM. Annual operation and maintenance.

APPA. American Public Power Association.

Array. Group of wind turbines sited at one location.

Articulating Blade. An airfoil that optimizes its angle of attack to the wind as the rotor turns.

ASME. American Society of Mechanical Engineers.

Average Windspeed. The mean windspeed over a specified period of time.

Avoided Cost. The sum of fixed costs (e.g., power plant, transmission lines) and variable costs (e.g., fuel) avoided by the utility when a dispersed power generator contributes energy to the utility grid.

AWEA. American Wind Energy Association.

Bearingless Rotor. A rotor in which the blades are attached to the hub without the need for bearings. A variable pitch rotor which depends on torsion to allow change in pitch of blades.

Betz Limit. The maximum efficiency theoretically obtainable from a WECS rotor, established by German scientist A. Betz as 16/27 (0.593) of the total kinetic wind energy within a given capture area and a given windspeed.

Blade. Element of a WECS rotor which forms an aerodynamic surface to extract energy from the wind.

Blade Tip. Portion of a blade farthest from the rotor axis.

BPA. Bonneville Power Administration

British Thermal Unit (BTU). A measurement of heat. The amount needed to raise the temperature of a pound of water one degree Fahrenheit.

BuRec. Bureau of Reclamation.

Busbar Price. The price of electricity at a generating plant, excluding the increment from the cost of transmission and distribution.

Buyback Rate. The rate per kilowatt hour (kWh) which a utility pays for excess energy fed into its lines by an outside generator such as a wind turbine.

Capacity Credit. A credit earned for ability to replace a conventional generating unit.

Capacity Factor. The actual amount of electricity generated by a power plant during a time interval, divided by the amount of electricity that would be generated by the plant during the same interval if it operated at maximum capacity.

Capital Costs. Investment costs required to build a system or device.

Capture Area. The maximum projected area of the rotor exposed to the wind.

Chord. Straight-line distance from the leading edge to the trailing edge of an airfoil.

COE. Levelized cost of energy (¢/kWh).

Coefficient of Performance. The ratio of power extracted by a WECS to the power available in the wind.

Coning. The downwind tilting of the blades from the hub. In a downwind turbine, coning protects the blades from excessive stress during high or gusty winds.

Coning Angle. The angle between a vertical plane and the rotor blades of downwind HAWTs.

Control System. WECS subsystem that senses the condition of the WECS or environmental parameters and adjusts WECS operation to protect it or optimize output.

Current. Flow of electricity through wires or other conduits.

Cut-in Windspeed. Windspeed at which a WECS begins to produce usable power. Not the same as start-up speed, which is the speed required to begin blade rotation. Care should be taken to define whether the cut-in speed is expressed at hub height or at some reference level.

Cut-out Windspeed. The highest windspeed above which a wind turbine produces no power. Care should be taken to define whether the cut-out speed is expressed at hub height or at some reference level.

Darrieus Machine. A vertical axis wind machine that has long, thin blades in a troposkein (skip rope) shape connected at the top and bottom of the axis; often called "eggbeater" because of its appearance.

DC. Direct current

Dedicated Storage. An energy storage system charged solely from WECS or any single energy source.

Demand. The amount of power required to satisfy the energy needs at a specific time.

Demand Charge. Charge by a utility based on the ratio of peak demand to total power consumed. It compensates the utility for maintaining sufficient generating capacity to meet a customer's immediate demand at all times.

Direct Current (DC). Electric current that flows in one direction, unlike alternating current (AC) which flows backward and forward.

Dispersed Systems. Term used to describe applications in which WECS are installed at many separate locations.

DOD. Department of Defense.

DOE. Department of Energy.

DOI. Department of Interior.

Downwind. On the side away from the wind. A wind turbine whose rotor is on the downwind or lee side of the tower.

Drag. A force which "slows down" the motion of wind turbine blades.

Drag Device. One of two major types of WECS. The other is a lift device. In a drag type, such as a Savonius rotor, the blades move in the direction of the wind. Generally, drag devices are less efficient at extracting the wind's energy than lift types and are more material-intensive.

Drive Train. WECS subsystem which transmits torque between the rotor and the generator.

Efficiency. A number arrived at by dividing the power output of a device by the power input to that device (usually the larger of the two numbers); usually expressed as a percentage value. See also *power coefficient.*

Eggbeater. Colloquial term for Darrieus rotors.

Electric Power Research Institute. Located in Palo Alto, California; the principal research arm of U.S. electric utilities.

Energy. The amount of work done over a given period of time. Wind energy is usually expressed in mechanical terms as horsepower hours (hph) or electrically in kilowatt hours (kWh). Not the same as power.

Energy Density. The amount of energy flowing in a windstream divided by the cross-sectional area of the windstream.

Energy Research and Development Administration (ERDA). Predecessor (before October, 1977) to the Department of Energy. Succeeded the Atomic Energy Commission in January, 1975.

EPA. Environmental Protection Agency.

EPRI. Electric Power Research Institute.

ERDA. Energy Research and Development Administration.

Fan. Colloquial term for rotor on American farm windmills.

Feather. To rotate the blade until it is parallel to the wind. By decreasing the area of blade surface exposed to the wind, feathering reduces thrust and stress on the rotor and tower.

Federal Energy Administration. Absorbed into the Department of Energy in October, 1977. Previously responsible for solar commercialization.

FERC. Federal Energy Regulatory Commission.

Firm Power. Power intended to have assured availability to the consumer to meet all, or any agreed-on portion, of his load requirements.

Fuel. A substance such as oil, coal, gas or wind that can be used to produce power.

Gear Ratio. The ratio of speeds (revolutions per minute) between the rotor shaft and the power shaft of a generator.

Gigawatt (GW). 1,000 million watts.

Gin Pole. A pole, pipe, or board used for leverage when raising or lowering a tower.

Governor. A mechanism that automatically controls the speed of a rotor, usually by changing the angle of the blades.

HAWT. Horizontal axis wind turbine.

Head. The height of water in a reservoir.

Height Diameter Ratio. For a VAWT, ratio of maximum vertical dimension to maximum horizontal dimension of the rotor.

Hertz (Hz). Unit of frequency equaling 1 cycle per second.

High Head. Refers to generation of hydroelectric power using large dams.

Horizontal Axis Wind Turbine (HAWT). A wind turbine whose main shaft is horizontal to the ground.

Horsepower (hp). A measure of power capacity.

Horsepower Hours (hp h). A measure of energy.

Hp. Horsepower.

Hub. The structural tie between HAWT rotor blades and drive train.

Hydroelectricity. The conversion of the kinetic energy in moving water (generally first held behind a dam) to mechanical (rotory) energy and then to electricity by a generator.

Hydrostorage. Technique to store power utilizing a dam by pumping water into a reservoir, to be drawn out when power needs to be generated.

IEEE. Institute of Electrical and Electronics Engineers.

Installed capacity. The total of the rated power of the units in a WECS facility.

Institute of Electrical and Electronics Engineers (IEEE). A major professional organization. Publishes literature and standards on a wide range of electricity generating devices.

Intercept Area. The area of the wind swept by a rotor. Also known as *swept area.*

Interface. The point at which wind-derived power mixes with regular, utility-supplied power. Used as a verb it means to interconnect wind-derived power with that of the electric grid.

Inverter. A device to convert direct current (DC) to alternating current (AC). See also *synchronous inverter.*

IOU. Investor owned utilities.

Kilowatt (kW). Unit of electricity equaling 1,000 watts of power; one hp equals 776 W or 0.776 kW.

Kilowatt Hour (kWh). An energy measure equal to the use of 1,000 watts for one hour.

kV. Kilovolt, a measure of electrical potential or potential difference, equal to 1,000 volts.

kVA. One thousand volt-amperes, a measure of power capacity.

kW. Kilowatt.

kWh. Kilowatt Hour.

Life Cycle Cost. A measure of what something will cost totally over its lifespan. The accumulation generally includes a discounting of future costs to reflect the relative value of money over time.

Lift. The force which "pulls" a wind turbine blade along, as opposed to drag.

Lift-type Devices. Devices that use airfoils or other types of shapes that provide aerodynamic lift in a windstream.

Load Factor. The average power output of a WECS divided by its rated power.

Low Head. Refers to generation of hydroelectric power with relatively small dams.

LWECS. Large wind energy conversion systems.

Marginal Cost. The peak cost for a utility to produce the next unit of electricity.

Maximum Design Windspeed. The highest windspeed which a turbine is designed to withstand intact.

Mean Windspeed. Arithmetic average windspeed over a specified time period and at a specified height above ground level. The international (WMO) standard 10m level may be implied.

Median Windspeed. The 50 percentile (i.e. 50% probably) windspeed value.

Megawatt (MW). A measure of power, equal to 1,000,000 watts (W) or 1,000 kilowatts (kW).

Megawatt Hour (MWh). Energy of 1,000,000 watts for one hour.

Meters/Second (m/s). Unit of metric speed. One meter/second is equal to 2.24 mph.

mph. Miles per hour.

m/s. Meters per second.

MW. Megawatt.

MWh. Megawatt-hour.

Nacelle. The covering that houses the generator and transmission assembly in horizontal axis wind turbines.

NASA. National Aeronautics and Space Administration.

NOAA. National Oceanic and Atmospheric Administration.

Normal Operations. The unattended operation of wind energy systems after installation and debugging.

NRECA. National Rural Electric Cooperative Association.

NSF. National Science Foundation.

NTIS. National Technical Information Service.

O&M. Operation and maintenance.

Off Peak. Refers to utility load demand or power generation occurring at other than peak load hours of the day.

Operation and Maintenance Costs. Those costs incurred after start-up of normal operation associated with maintaining a wind energy system so that it continues to perform satisfactorily over its design lifetime.

Overspeed Control. Means by which the speed of a rotor is controlled during periods of high wind.

Panemone. From the Greek pan (all) anemone (winds). Omnidirectional.

Parallel Generation. The generation of electricity from a WECS interconnected with the utility grid.

Payback. A traditional measure of economic viability of investment projects. A payback period is defined in several ways, one of which is the number of years required to accumulate fuel savings which equal the initial capital cost of the system.

Peak Generating Capacity. The maximum power output a wind system is capable of producing in normal operation.

Peaking Units. Utility generating units assigned solely to respond to the periods of highest load demand.

Pitch. The angle of the blade, measured between the chord line and the direction of motion.

Plant Factor. The actual energy generated by a wind turbine divided by the potential energy available if the wind turbine operated at its rated power 100 percent of the time.

PNL. Pacific Northwest Laboratory.

Post Mill. An early type of European windmill which turned about a central post. To turn it into the wind the miller turned the mill by means of a long pole known as a *tail pole.*

Power. The capacity or rate at which work is performed. Usually measured mechanically in horsepower (hp), electrically in watts (W), or calorically in British thermal units (BTUs) per hour. 1 hp is equivalent to 746 W. 1 W is equivalent to 3.413 BTUs.

Power Coefficient. The ratio of power extracted by a WECS to that available in the capture area of the free windstream. The ratio of power output to power input.

Power Conditioning. Converting electrical energy generated by a wind turbine into other electric forms, such as DC to AC.

Power Curve. A plot of WECS power output versus windspeed.

Power Density. The amount of power per unit of cross-sectional area of a windstream. Usually given in watts/square meter (W/m^2).

Power Duration Curve. A graph of power generated by a WECS vs. time.

Power Output. Useful power generated by the WECS at any given time.

Prevailing Wind. The wind occurring most frequently at a site but not necessarily providing the most energy. Because the power in the wind increases as the cube of windspeed, a stronger but less frequent wind may contain more wind energy. This is an important distinction to be noted when siting a wind turbine.

Public Utility Regulatory Policies Act (PURPA). Bans discrimination against small power producers and directs utility purchase of small power production at rates commensurate with the utility's avoided (or marginal) cost.

PUC. Public Utility Commission.

PURPA. Public Utility Regulatory Policies Act.

Quad. One quadrillion (10^{15}) BTU. Commonly used as measure of annual energy consumption, usually expressed as primary fuel equivalent. Present U.S. consumption is about 78 quads.

Rated Power. The power output (kW) of a wind turbine can be its maximum power, or a power output at some windspeed less than the maximum speed before governing controls reduce the power.

Rated Windspeed. The lowest windspeed at which the rated output power of a wind machine is produced.

Rayleigh Windspeed Distribution. A mathematical idealization giving a ratio of time the wind blows within a given windspeed band to the total time under consideration. This distribution is dependent only on mean windspeed.

R&D. Research and Development.

RD&D. Research, Development, and Demonstration.

REA. Rural Electrification Administration.

Reactive Power. The quantity in AC circuits that is obtained by taking the square root of the difference between the square of the kilovolt-amperes (kVA) and the square of the kW. It is expressed as reactive volt-amperes or vars.

REC. Rural Electric Cooperative.

Renewable Resources. Sources of energy that are regenerative or virtually inexhaustible, such as solar and wind energy.

RFP. Request for proposal.

Rotor. The power-producing blades and rotating components of a wind turbine.

Rotor Diameter. The distance from the center of rotation of a wind turbine to the outermost point of the blade multiplied by two. A term commonly used to describe the size of a WECS. In general, the larger the rotor diameter, the greater the amount of energy captured from the wind.

rpm. Revolutions per minute.

Sandia. Sandia National Laboratories.

Smock Mill. An early European windmill in which the cap containing rotor, main shaft, and gear assembly turned about the vertical axis of the tower.

Soft. Flexible. Term used to describe towers that flex under load, such as a tubular tower made from a steel pipe.

Solidity. The ratio of blade surface area to rotor swept area.

Spar. The main structural member in many blade designs.

SPP. Small Power Producer.

Stall. Loss of lift resulting from increasing angle of attack, causing the blades to stop or slow rotation.

Stand-Alone. A WECS designed for self-sufficiency without outside power.

Start-up Windspeed. Minimum windspeed at which a rotor at rest will begin to rotate consistently.

Survival Windspeed. Maximum windspeed a WECS can sustain without damage to structural components or loss of ability to function normally.

SWECS. Small wind energy conversion systems.

Swept Area. The area of the wind swept by a rotor. Also known as intercept or capture area.

Synchronous Inverter. A DC to AC inverter, the output of which is synchronized with an exterior AC power source.

System-Wide Storage. An energy storage system accessible to, and chargeable by, any generating source in the system having available and/or excess capacity.

Tip Speed. Speed of a rotor at the circumference of its path.

Tip Speed Ratio. The ratio of blade tip speed to windspeed. Modern WECS used high-speed airfoils that operate at tip-speed ratios of more than 5 : 1. Drag devices run at much lower tip-speed ratios.

Torque. A measure of force from the rotor causing the power shaft to turn. Power is the product of torque and shaft speed or rpm.

Tower. Support structure.

Tower Shadow. Wake created by airflow around a tower.

Transmission. In a WECS, a mechanical or hydraulic device for converting the low speed and high torque of the rotor to the high speed, low torque needed to generate electricity.

Troposkein. From the Greek for "skipping rope," the shape of the rotor in a Darrieus system.

Turbulence. Rapid fluctuations of windspeed.

TVA. Tennessee Valley Authority.

Upwind. On the same side as the direction from which the wind is blowing.

USDA. United States Department of Agriculture.

V. Volts.

Vane. Blade. Often refers to the curved metal blades of the American farm windmill.

Variable Pitch. A method of controlling rotor torque and speed where blade pitch is varied to produce the required effect.

VAWT. Vertical axis wind turbine.

Velocity Duration Curve. Graph of wind velocity and the amount of time the wind occurs at given speeds.

Vertical Axis Wind Turbine (VAWT). A wind turbine whose rotor axis is perpendicular to the ground.

Volt. A unit of electrical potential or potential difference.

Voltage (V). The electrical pressure which causes current flow (amps).

W. Watt.

Wake. Disturbed flow downwind of an obstruction.

Watt. A unit of electrical power; watts = volts $\times$ amperes.

Watt Hours (Wh). Unit of electric energy; see also *kilowatt-hours*.

Watts per Square Meter (W/m^2). A measure of the energy in the wind passing through a square meter of area.

WECS. Wind energy conversion system.

Windcharger. Small wind generators of the 1930s and 1940s used to charge batteries, later increased in size to provide home lighting and other electric needs.

Wind Energy Conversion System (WECS). A general term including all the hardware components (e.g. rotor, generator, power converter) that transform energy from the wind into usable form.

Windfarm. A group of WECS sited at one location producing power jointly.

Windfurnace. A wind-powered electrical resistance heating system used to heat water.

Windmill. Archaic term for wind system; still used to refer to high-solidity rotor waterpumpers and older mechanical output machines.

Windpower. Power in the wind, part of which can be extracted by a wind turbine. See also *power*.

Windpower Profile. How the windpower changes with height above the surface of the ground or water; the windpower profile is proportional to the cube of the windspeed profile. See *windspeed profile*.

Wind Regime. The sum of wind characteristics in a given location.

Wind Rose. Circular bar graph of windspeed, direction, and frequency of occurrence.

Windspeed Duration Curve. The cumulative probability curve for windspeed, expressed in probability units or in hours/year.

Windspeed profile. How the windspeed changes with height above the surface of the ground or water.

Wind Turbine, Wind System, or Wind Machine. Accepted modern terms for devices which extract power from the wind; can refer to devices which produce mechanical or electrical power output.

Wind Turbine Generator. A wind system which produces electrical power; sometimes abbreviated WTG. Preferred use is wind energy conversion system (WECS).

Wind Velocity. Windspeed.

WTG. Wind turbine generator.

Yaw. The pivoting of a HAWT to face into or away from the wind.

Appendix C
Sources of Information

U.S. FEDERAL WIND ENERGY PROGRAM

General Information

Conservation and Renewable Energy Inquiry and Referral Service
Renewable Energy Information
P.O. Box 1607
Rockville, MD 20850

U.S. Department of Energy (DOE)
Division of Solar Energy, Wind Systems Branch
600 E Street NW
Washington, DC 20545
   (202) 252-5000

Specific DOE Wind Programs:

Small Wind Systems
DOE Rocky Flats Wind Systems Program
P.O. Box 464
Golden, CO 80401
   (303) 497-7000

Large Horizontal Axis Wind Turbines
NASA-Lewis Research Center
Wind Power Office
21000 Brook Park Road
Cleveland, OH 44135
   (216) 433-4400

Siting and Wind Resources
Pacific Northwest Laboratories (PNL)
Battelle Boulevard
P.O. Box 999
Richland, WA 99352
   (509) 942-4410

Agriculture Applications
U.S. Department of Agriculture (USDA)
SW Great Plains Research Center
Bushland, TX 79012
  (806) 378-5734

Vertical Axis Wind Turbines
Sandia National Laboratories (Sandia)
Information Division 5712
Albuquerque, NM 78185
  (505) 844-3850

Specific Wind Data

Local Wind Data Summaries
Environmental Data Service
National Climatic Center
Ashville, NC 28801

OTHER U.S. GOVERNMENT INFORMATION

Department of Commerce (DOC)
National Technical Information Service
5285 Port Royal Road
Springfield, VA 22161
  (703) 557-4600

Department of the Interior (DOI)
Office of Public Affairs
18th and C streets NW
Washington, DC 20240
  (202) 343-3171

Federal Energy Regulatory Commission (FERC)
Public Inquiries Branch
Office of Congressional and Public Affairs
825 N. Capitol Street NE
Washington, DC 20426
  (202) 357-8055

National Energy Information Center
Office of Energy Information Services
Energy Information Administration
EI-72, Forrestal Building
Washington, DC 20585
  (202) 252-8800

Congressional Resources:

General Accounting Office (GAO)
441 G Street NW
Washington, DC 20548
(202) 275-6241

Office of Technology Assessment (OTA)
Public Communications Office
600 Pennsylvania Avenue SE
Washington, DC 20510
(202) 226-2115

U.S. House of Representatives

Energy and Commerce Committee
2125 Rayburn House Office Building
Washington, DC 20515
(202) 225-2927

Legislative Status Office
House Annex Building No. 2
2nd and D Streets SW
Washington, DC 20515
(202) 225-1772

Science and Technology Committee
2321 Rayburn House Office Building
Washington, DC 20515
(202) 225-6371

U.S. Senate

Energy and Natural Resources Committee
3104 Dirksen Senate Office Building
Washington, DC 20510
(202) 224-4971

U.S. TRADE ASSOCIATIONS

American Public Power Association (APPA)
2301 M Street NW
Washington, DC 20037
(202) 775-8300

American Wind Energy Association (AWEA)
1050 Seventeenth Street NW
Suite 1100
Washington, DC 20036
　(202) 775-8910

Edison Electric Institute (EEI)
1111 19th Street NW
Washington, DC 20036
　(202) 828-7600

Electric Power Research Institute (EPRI)
3412 Hillview Avenue, P.O. Box 10412
Palo Alto, CA 94303
　(415) 855-2872

National Rural Electric Cooperative Association (NRECA)
2000 Florida Avenue, NW
Washington, DC 20009
　(202) 857-9534

CANADIAN GOVERNMENT WECS PROGRAM

National Research Council of Canada (NRC)
Low-Speed Aerodynamics Laboratory, M-2
Montreal Road, Ottawa, Ontario KIA OR6
　(613) 993-9127

MAJOR U.S. STATE PROGRAMS

California

California Energy Commission
1111 Howe Avenue
Sacramento, CA 95825
　(916) 920-6031
　Rusty Schweickart

Hawaii

Hawaii Natural Energy Institute
Holmes Hall 240
2540 Dole Street
Honolulu, HA 96822
　(808) 948-8788
　Dick Neill

Kansas

Kansas Energy Office
214 West Sixth St.
Topeka, KS 66063
(913) 296-2496

Oregon

Oregon Department of Energy
102 Labor & Industries Bldg.
Salem, Oregon 97310
(503) 378-6715
Donald Bain

Texas

TENRAC
200 East 18th St., 5th Floor
Austin, TX 78701
(512) 475-5588
Bob Avant

NORTH AMERICAN WECS PRODUCERS

Aluminum Company of America (Alcoa)
1501 Alcoa Building
Pittsburgh, PA 15219
(412) 553-4545
Frank Townsend

Bendix Corporation
2582 South Tejon Street
Englewood, CO 80110
(303) 922-6394
Fred Whitson

Boeing Engineering & Construction
Div. of the Boeing Company
P.O. Box 3707
Seattle, WA 98124
(206) 575-5922
John Lowe

California Energy Group, Inc.
3500 South Susan Street
Santa Ana, CA 92704
(714) 966-1202
Rick Carroll

Carter Wind Systems
P.O. Box 684
Burkburnett, TX 76354
(817) 569-2238
Jay Carter, Jr.

DAF-Indal Ltd.
3570 Hawkestone Road
Mississauga, Ontario, Canada L5C 2V8
(416) 275-5300
Chuck Wood

Energy Sciences, Inc.
P.O. Box 3168
Boulder, CO 80307
(303) 449-3559
Jim Alexander

Enertech Corporation
P.O. Box 420
Norwich, VT 05055
(802) 649-1145
Bob Sherwin

Flow Industries, Inc.
21414 68th Avenue South
Kent, WA 98031
(206) 872-8500
Tom Hiester

Forecast Industries, Inc.
3500A Indian School Road NE
Albuquerque, NM 87106
(505) 265-3707
Paul Vosburgh

General Electric Company
P.O. Box 8661
Philadelphia, PA 19101
  (215) 962-1219
  Dick Calef

Hamilton Standard Division
Wind Energy Systems Dept.
Mail Stop 1-3-8
Windsor Locks, CT 06096
  (203) 623-1621
  Bob Gregoire

PM Wind Power, Inc.
P.O. Box 89
7050 Maple Street
Mentor, OH 44060
  (216) 255-3437
  John Deering

WECS-Tech Corporation
4327 Redondo Beach Blvd.
Lawndale, CA 90260
  (213) 542-1666
  Greg Crist/Al Hunter

Westinghouse Electric Corporation
875 Greentree Road
Bldg. 8, 4th Floor
Pittsburgh, PA 15236
  (412) 928-2430
  Will Treese

Wind Engineering Corporation
P.O. Box 5936
Lubbock, TX 79417
  (806) 763-3182
  Coy Harris

Wind Power Systems, Inc.
8630 Production Avenue
San Diego, CA 92121
  (714) 566-1807
  Ed Salter

Windtech, Inc.
P.O. Box 837
Glastonbury, CT 06033-0837
    (203) 727-7536 or (203) 633-7582
    Kip Cheney

WTG Energy Systems, Inc.
251 Elm Street
Buffalo, NY 14203
    (716) 856-1620
    Al Gross

CONSULTANTS, WRITERS, AND ADVOCATES

AeroVironment, Inc.
145 North Vista Avenue
Pasadena, CA 91107
    (213) 449-4392
    Peter Lissaman

Alternative Energy Consultant
AFS
Rt. 2, Box 2672C
LaGrande, Oregon 97850
    (503) 963-3362 or (503) 963-7961
    Glen Andrews

Alternative Energy Institute
West Texas State University
Box 248
Canyon, TX 79016
    (806) 656-3904
    Vaughn Nelson

Alternative Energy Resources, Ltd.
201 E. 10th Street
New York, NY 10003
    (212) 475-3554
    Bill Warburton

The Center for Alternative Resources
P.O. Box 539
Harrisburg, PA 17108
    (717) 238-5214
    Paul Gipe

Clean Energy Products
3534 Bagley North
Seattle, WA 98103
  (206) 633-5506
  Ed Kennell

Engineering Research Institute
University of New Mexico
Box 25, University Station
Albuquerque, NM 87131
  (505) 844-6187
  Gerry Leigh

Energy Systems, Inc.
3901 W. International Airport Road
P.O. Box 6065
Anchorage, Alaska 99502
  (907) 243-1942
  Bill/Mark Ogle

Arthur D. Little, Inc.
Acorn Park
Cambridge, Mass. 02140
  (617) 864-5770
  Bill Vachon

Managerial Controls, Inc.
5910 Merritt Place
Falls Church, VA 22041
  (703) 379-7734
  Harry Cruver

Natural Power, Inc.
New Boston, NH 03070
  (603) 487-5512
  Rick Katzenberg

Strategies Unlimited
201 San Antonio Circle
Suite 205
Mountain View, CA 94040
  (415) 941-3438
  Robin Sacks

Sunflower Power Company
Box 862
Lawrence, KS 66044
   (913) 841-1906
   Steve Blake

The Synectics Group, Inc.
1120 19th Street NW
Washington, DC 20036
   (202) 887-0970
   Mike Lotker

Transition Energy Projects Institute
1045 Broadway
San Francisco, CA 94133
   (415) 474-9463
   Randy Tinkerman

Wind Energy Report
Box 14 – 104 S. Village Avenue
Rockville Centre, NY 11571
   (516) 678-1230
   Farrell Seiler

Index

Index